Our world is in desperate need of leader... leadership that provides comfort to those ~~.. ......~~ and depleted after years of political rancor, and a path forward that inspires and empowers us all to take action and drive change. What Michael Slaby has put forth in *For ALL the People* confirms his status as one of those leaders. The book combines the big thinking that is too often lacking from today's political dialogue and action steps that will actually make a difference. To say it is a must-read is an understatement.

**—BRIAN REICH, AUTHOR OF *THE IMAGINATION GAP* AND VP OF COMMUNICATIONS, MURMURATION**

Slaby deftly illuminates our agency for taking back control of our community and leadership narratives, highlighting the ways that we can and must change antiquated media frameworks, public policy, legislation, and business models that govern our daily storytelling. In order to benefit from the promise of infinite digital connection, he exhorts us to create a modern public and technology-enabled space where not only information, but also experience is freely shared, authenticated, and given credibility. From a business perspective, this resonates as someone who believes that empathy is currency—creating beautiful and useful products can only happen through intense listening, openness to feedback, and focus on purpose. This book serves as an awakening for innovators seeking to drive positive social change and a reclamation of our democracy.

**—ELIZABETH BRIGHAM, DIRECTOR, HURT HUB@DAVIDSON**

*For ALL the People* is utterly necessary and timely. Slaby illuminates the how and why driving the existential dysfunction of our media ecosystem, and offers an incisive, thorough prescription toward something better. Essential reading for every storyteller, politician, technologist, and citizen.

**—LYEL RESNER, PROFESSOR, NYU INTERACTIVE TELECOMMUNICATIONS PROGRAM**

This is required reading for anyone working in media, technology, or civic life. This thoughtful dissection of modern media and its impact on democracy has implications for both public and private sector leaders. Slaby paints a picture of a new reality where civic conversation is productive and spurs on innovation. With clear, incremental steps to guide the construction of these new spaces, a healthy civic dialogue is in reach for all Americans.

**—ALEX JOHNSTON, FOUNDER, CITIES REIMAGINED**

Michael Slaby's *For ALL the People* should be required reading for anyone who watches the news or engages in social media—so pretty much all of us. Slaby is the Yuval Noah Harari of media. He pointedly outlines the fracture of our culture and democracy while providing hopeful next steps for how we can redesign our systems for a more civically minded world.

**—MEGAN KIEFER, AUTHOR OF *THE THIRD DIMENSION OF LITERACY* AND FOUNDER, TAKE TWO FILM ACADEMY**

My greatest concern about the increased division and hyperpartisanship of the last decade is the unfounded belief that it is happening to us, not by us. The existential threat in that belief is the lack of agency to fix it. *For ALL the People* does two critical things—it helps us understand how we got here, and it reminds us that we have the agency to repair it. Perhaps most important and prophetic, Slaby shows that indeed we are the only ones who can.

**—JASON GREEN, FORMER ASSOCIATE WHITE HOUSE COUNSEL AND DIRECTOR, FINDING FELLOWSHIP**

*For ALL the People* examines how technology has rapidly and fundamentally changed how we engage in civil discourse, how we interact with governing institutions, and what can happen when systems are manipulated by those seeking to erode the pillars of a democratic society. After four years of our spiraling through disinformation cycles and infodemics, Slaby puts forward important ideas that can move us toward restoring trust in the information we receive and will enable a return to progress.

**—LISA KAPLAN, FOUNDER AND CEO, ALETHEA GROUP**

This book is for anyone who has thought to themselves, 'It doesn't have to be this way.' Michael Slaby offers a vivid reimagining of civic life in America, one in which technology and media bring us together instead of tearing us apart; one in which we unlearn the habits ingrained in us by a media economy optimized to capture our attention by any means necessary. Slaby shows us that it is possible to abandon the zero-sum mentality we've been conditioned to embrace, and reminds us that it is up to us to create the culture we wish we had—one born of community, dialogue, and the stories we tell.

—KATE CATHERALL, FOUNDER OF ARENA AND EVP OF
POLITICAL STRATEGY, AVALANCHE INSIGHTS

In *For ALL the People*, Slaby indeed writes for all the people: the trials and truths, the prescriptions and possibilities he outlines are universally applicable to people and societies everywhere. Indeed, the fraying of American culture was presaged by the fraying of polities in other parts of the world. But, it then also holds that the commitments Slaby insists we as Americans can make today, both individually and institutionally, should give hope to people everywhere, too. For if the problems of design, discourse, and democracy that have swept our national culture in recent years are, in fact, currents that travel far beyond our shores, then his insistence that we redesign, reclaim, restore, and then, finally, redeem the shared experiences that constitute our modern media should awaken all Americans to our opportunity to set an example for all the world.

—RISHI JAITLY, FOUNDING CEO, TIMES BRIDGE;
FORMER VICE PRESIDENT, ASIA PACIFIC/MIDDLE EAST,
TWITTER; AND FORMER GOOGLE/YOUTUBE

In *For ALL the People*, Michael Slaby shows us how to think anew about media and the tsunami of information coming at us in our age of technology. This is a book for all of us, offering insights and actions to help make sense of our new world and connect more powerfully with one another.

—BETH COMSTOCK, AUTHOR OF *IMAGINE IT FORWARD*
AND FORMER VICE CHAIR, GE

In a world dominated by hot takes and clever sarcasm, Slaby offers us a new kind of conversation about how to leverage innovation and reclaim politics. An earnest and optimistic must-read that can help lead us to a better place as a nation.

—LAUREN ZALAZNICK, FORMER CHAIR, ENTERTAINMENT AND DIGITAL NETWORKS, NBCUNIVERSAL

Nothing is easier than blaming social media for our broken national dialogue. If only it was that simple. How we communicate with one another is both symptom and cause. Michael Slaby does stellar work explaining how we got to this point as a society and, even more important, presents a compelling—and constructive—vision of where we go from here.

—IAN BREMMER, FOUNDER AND PRESIDENT, EURASIA GROUP

Michael Slaby offers a prescription to cure our social media ills, from redesigning the platforms to how we reclaim our own voices and stories. It's a must-read for anyone interested in disrupting the toxic social echo chambers and bringing about positive change.

—CHARLENE LI, *NEW YORK TIMES* BESTSELLING AUTHOR OF *THE DISRUPTION MINDSET*

# FOR *ALL*
# THE
# PEOPLE

# FOR ALL THE PEOPLE

REDEEMING the BROKEN PROMISES
of MODERN MEDIA and
RECLAIMING OUR CIVIC LIFE

## MICHAEL SLABY

DISRUPTION
BOOKS

AUSTIN    NEW YORK

Published by Disruption Books
New York, NY
www.disruptionbooks.com

Distributed by Disruption Books

For ordering information or special discounts for bulk purchases, please contact
Disruption Books at info@disruptionbooks.com.
Cover and text design by Kimberly Lance
Library of Congress Control Number: 2020923053

Print ISBN: 978-1-63331-051-3
eBook ISBN: 978-1-63331-052-0

First Edition

*For Lydia, without whom there would be little reason to write.*
*And Ms. Denizé, without whom I would not know how.*

# CONTENTS

**FOREWORD**: Our Story, by Deval Patrick . . . . . . . . . . . . . . . . . . . .xiii

**INTRODUCTION**: We Were Promised Jetpacks. . . . . . . . . . . . . . . . . 1

**1**: Lost Like a Goat in a Pharmacy . . . . . . . . . . . . . . . . . . . . . . . 13

**2**: I Was Told There Would Be No Math on This Exam . . . . . . . . 33

**3**: It's Not the Crime—It's the Coverup . . . . . . . . . . . . . . . . . . . 63

**4**: You May Ask Yourself, "How Did I Get Here?". . . . . . . . . . . . 83

**WHERE DO WE GO FROM HERE?** . . . . . . . . . . . . . . . . . . . . . . 101

**5**: Redesign Our Platforms . . . . . . . . . . . . . . . . . . . . . . . . . . . . 105

**6**: Reclaim Our Voices . . . . . . . . . . . . . . . . . . . . . . . . . . . . . . . . 127

**7**: Restore Our Institutions . . . . . . . . . . . . . . . . . . . . . . . . . . . . 143

**8**: Redeem Ourselves . . . . . . . . . . . . . . . . . . . . . . . . . . . . . . . . . 161

**EPILOGUE**: Manufacturing Outrage. . . . . . . . . . . . . . . . . . . . . . 175

*References* . . . . . . . . . . . . . . . . . . . . . . . . . . . . . . . . . . . . . . . . . . 183

*Acknowledgments* . . . . . . . . . . . . . . . . . . . . . . . . . . . . . . . . . . . . 187

# OUR STORY

### *by Deval Patrick*

M Y STORY BEGAN on the South Side of Chicago. Growing up in my grandparents' two-bedroom tenement, I shared a room with my mom and my sister, rotating between top bunk, bottom bunk, and floor every third night. My grandmother told us that we weren't poor, just broke—because broke is temporary.

My grandmother (like most Southerners of her vintage, it seems) knew that the story we told ourselves about who we were and the lives we were living mattered. *Poor*—at least then, before America lapsed into associating poverty with the unrelated concept of fault—told a story about deprivation. *Broke* was about a state of affairs, not a state of being. It left room for aspiration and expectation. My grandmother meant for us to see ourselves in motion, becoming something, shaping our own future.

Decades later, when I was governor of Massachusetts, the power of the stories we told each other were a part of leadership. In the wake of the Boston Marathon bombing in 2013, for example, telling the stories of the many acts of kindness done by residents for runners, visitors, and other strangers seemed to inspire more of the same; individual generosity and grace took on a value as important as that of the work of first responders and medical professionals. And stories from constituents about their long, rather ordinary lives with same-sex partners, told by them directly to their representatives in the state legislature, made it possible to win the legislative fight for marriage equality—making Massachusetts the first state in the nation to do so.

The American Dream has been an important story for me, both personally and as a policy maker. It speaks to our ability to imagine a better way for ourselves and our families—and to reach for it. In my hearing and telling, it evokes a sense of community, where we share in each other's striving. It is unifying. For those of us governing through the Great Recession, restoring the American Dream was the goal we shared as a nation. It was the reason for the choices and the public investments we made.

Sadly, like too many of our shared stories in recent years, the American Dream has been hijacked by unhealthy trends in media.

Michael Slaby has been thinking about the impact of the stories we tell one another since his days as a leader on the 2008 presidential campaign of Barack Obama—where he was one of the geniuses behind the campaign's then-revolutionary digital strategy—when I first came to know him. His storytelling, like his candidate's, spoke a promising and inclusive language about enabling the unseen and unheard to be seen and heard at last. This language, coupled with the capacity of social media to share such stories, offered a political advantage that was instantly clear. It was even more compelling as an expression of and a tool for the kind of servant leadership that Obama and I both still believe in.

Of course, the value and power of stories depends on who tells them, what those storytellers intend, and which community gets to hear them. If the storyteller is fake, the intent is sour, and the community is designed in such a way that it hears nothing else, the impact can be sinister.

In today's world, information has been weaponized. Our data and our lives have become the inventory of the platforms we rely on to learn, to connect, and to communicate. Shared storytelling is compromised. Inside systems that sort us into separate, addressable audiences for advertisers, stories meant to unite us don't resonate with the same intensity. Propaganda flourishes. Platforms ostensibly meant to build community are actually narrowing our sense of what that means. When we are more isolated from others—especially from others whose life experiences aren't like our own—we suffer from a world less complete, a national community less united, and people less connected to and invested in each other's success.

In *For ALL the People*, Slaby brings us a jeremiad. He takes us on a dark journey through the unkept promises and unappreciated hazards of the digital revolution. The speed with which information and stories emerge, evolve, and demand our attention has changed our expectations of ourselves as individuals and as a society. That insistence on speed, on immediate gratification, on short-term results and quarterly gains is marbled throughout our civic life. Simultaneously, our ability to unite in common cause has eroded. When did we lose our sense of opportunity, of steadfastness, of commitment to one another? When did outrage become the emotion of choice? Where are our better values?

This loss is not random, Slaby argues, nor is it the inevitable path of democracy in the age of social media. It is the consequence of how we interact, where we interact, and the norms those systems create in us that we then bring to our civic life. And here is where Slaby is both prescriptive and hopeful.

Slaby urges in these pages, first, that we should care about what habits our media and information systems are creating in us, and then, that we should ask, *Who benefits?* We need to understand the consequences for our ability to remain in community, in the massive, messy, imperfect union that is America in an era when we lack a shared idea of what that union means.

In my story, all of the adults on the block took responsibility for all the kids. I knew that if I messed up down the street, I was going to get a smack from Mrs. Jones and then another when I got home. Every child was under the jurisdiction of every adult because we were invested in one other, and so we created a story together, every day, about who we were and how we were bound to each other. Membership in our community meant understanding the stake we had in our neighbors' dreams as well as our own, and that meant holding each other to account with love and commitment. Our shared stories told us who our neighbors were. And in learning about them, we learned about ourselves.

The American Dream is meant to be a shared story—a story of shared success, shared opportunity—and it works only when we are invested in each other's shared striving and thriving. It was flawed from the start, of course: too white, too wealth-centric, and too male to be truly complete. For many, it has often represented not what is possible but what is withheld. I do not accept that the American Dream is about greed or self-centeredness. But as the systems we use to tell and to consume our stories have evolved, we find ourselves unwilling or unable to listen to a telling of the American Dream that is about meaning and purpose.

I want to believe that greater connectivity can lead to greater connection and to greater diversity of people and perspectives. I want to believe that the wild innovation of tools and networks that have given rise to new voices can lead us to better understanding, better vision, and more investment in our neighbors (both digital and physical). I want to believe that digital media is about more than using or beating others in a zero-sum competition for wealth and achievement—that these tools can live up to their vaunted promise. I hope Slaby is right.

I hope we can reclaim and revive the story of the American Dream as *our story*, about shared progress for all of us.

# WE WERE
# PROMISED JETPACKS

THE MORNING OF November 5, 2008, I stumbled from an anony-
mous office building in Chicago's Loop into the chilly, brilliant sun-
shine of an America that had just elected Barack Obama its first Black
president. The morning after the Grant Park Election Night celebration,
our team at Obama for America had to launch the Change.gov transition website, so I was bleary-eyed from yet another all-nighter. I had that
semi-nauseated feeling you get as adrenaline wears off and was mildly
incoherent from a strange mix of the elation of winning mingled with the
letdown of crossing the finish line only to discover miles more to go.

It was an unquestionably historic moment, the culmination of tens of
millions of hours of work done by millions of people who were involved
in the campaign, and both the best and worst job I'd ever had. It felt like
a moment of triumph and a moment of political upheaval. President
Obama's campaign had been driven by people transcending stereotypes
and embracing that politics was about us, not about him—about our better
angels, not our lesser ones. For the first time in my lifetime, I had the faint
but durable feeling that progress was not just possible but inevitable.

But that historic moment reverberated very differently across the
country, experienced in diverging landscapes: the high-fiving on the left

was accompanied by despair on the right. It was the beginning of a new wave of increasingly partisan conflict that would spawn the conservative Tea Party movement and accelerate the further estrangement of American political life, inflamed by systems of information that were supposed to be making our world more interconnected.

Central both to that victory and to that intensifying estrangement were modern media systems that helped us discover and empower a whole new electorate. We had found ourselves in a new, increasingly digital world that created new opportunities to embrace storytelling in dramatically expanded ways. We had built an entirely new social network just for our supporters, harnessing the uncontrollable chaos of social media to help cultivate an intense culture of organizing. Digital tools had helped breathe life into a vibrant Democratic electorate and provided the foundation for a new quantum leap in the potential of distributed organizing and grass-roots fundraising.

We did not invent organizing. The basic structure of campaigning is largely unchanged since President McKinley's campaign manager, Mark Hanna, invented it in 1896: raise money; deliver messages; mobilize voters. But bringing campaigning into a new, increasingly digital age presaged a massive shift—a proverbial Rubicon that we cannot go back across culturally. In this new world, our experiences not just of politics but of culture and history—of a generational, historic moment—are not shared in the same way that William Jennings Bryan's Cross of Gold speech (during that same campaign against President McKinley in 1896) shifted American ideas about economics for almost everyone at once. That shared experience did not demand and did not suggest unified opinion. In 1896 the election was largely decided on the conflict between bimetallism and the gold standard. But Bryan's speech provided a shared cultural moment, a shared foundation of experience that bound together a country despite profound disagreement.

In 1996, Elizabeth Corcoran wrote in the *Washington Post* that "the Web is a crazy quilt of both utopian and Orwellian possibilities." That simple sentence not only acknowledges the cyberutopian promise of greater connectivity inexorably leading to a more vibrant, productive Western liberal democracy, but also delivers a warning against a new form of cultural authoritarianism in the form of the tyranny of outrage. The internet's original creators themselves recognized its inherent disruptiveness. And Elizabeth Eisenstein, historian of movable type and the printed word, in 2005 likened it to the printing press in its capacity to fundamentally shift our paradigms of knowledge, storytelling, and relationships, and in turn to interrupt standards and expectations for civic and economic life.

They weren't wrong.

The initial research for the precursor to what we now experience as the internet was funded by the US Department of Defense through its Defense Advanced Research Projects Agency (better known as DARPA). In those heady, academic early days of design and creation, of open protocols and universal access, the engine of commercialization that would drive the mass expansion of the internet and reorder our basic landscape of information was still beyond the horizon. Unseen, it was unregarded and unplanned for—and thus we left our most important needs and desires surrounding how a new architecture of information would reshape society to be *implicit* assumptions about the inevitability of greater connectivity, greater diversity of voice, and greater distribution of access and power. Without *explicit* direction, commercial interests have optimized modern media for profit, not for civic life or human progress, and our civic life is collapsing under the weight of exploitation.

The future we had hoped for—a more informed, more open, more equal, more connected society—has yielded instead a never-ending stream of history-less reactions to twenty-second clips of teenage beatdowns, celebrity gaffes, and stupid (albeit hilarious) pet tricks. We were promised jetpacks, but we got cat videos.

**THE 2008 PRESIDENTIAL** campaign was full of both obvious firsts, like our first Black president, and less obvious ones among the dozens of digital firsts that the Obama campaign pioneered on the way to victory: first Facebook page, first Twitter account, first this, first that. There is an obvious truth about innovation in politics: each cycle is the most technologically advanced cycle of all time—and this was especially true in 2008. Many of the firsts attributed to Obama for America were not available to anyone before us. Facebook and Twitter weren't options for the Kerry or Bush campaigns in 2004. "Timing has a lot to do with the success of a rain dance," as venture capitalist Chris Sacca is fond of saying.

We waged a massive national campaign on the heels of the initial explosion of today's ubiquitous media giants: only eighteen months after Facebook opened up to users outside of college communities and mere months after the launch of Twitter. We were a campaign with an energetic young voice calling for dramatic generational change that was destined to lose badly in a traditional Democratic primary. A traditional Democratic primary electorate was always going to vote for a traditional Democratic candidate, and that candidate certainly wasn't the then-senator from Illinois with the funny name that no one had really heard of. The only way to win the 2008 Democratic primary was to change who participated in it.

The innovation that the Obama for America campaign has been so lauded for was driven by political desperation, not virtue. We were desperate for new ways to reach, inspire, and organize new people in a political process that had generally ignored or outright excluded them. Having little experience and nothing to lose meant we had yet to subscribe to hardened, safe thinking. We were an organization of nontraditional thinkers unbound by conventional wisdom in need of new ground to win.

Coming into that 2008 campaign, we all felt the early effects of social media knitting back together all the fragments of the 1990s and early 2000s. The breakdown of large, stable audiences was the norm by then, but new habits for how to leverage this new world weren't yet established. Desktop publishing had reached a zenith where prosumer production was very nearly the same quality and capacity as actual professional production. People had begun to develop greater expectations of what it meant to be invited into a movement online, but they hadn't become cynical yet about

movement-building as a marketing technique. The campaign had no existing infrastructure to bring to the table—no real inertia to hold us back (or
keep us centered)—but we knew that people talking to other people, orga-
nizers helping people build real power in their communities, was going to
be the key to a whole new movement cut out of entirely new cloth.

We began to see all of these interactions—online and off, storytelling and
organizing—as one big network, where people were powerfully connected
in an ever-expanding number of ways that could determine which messages
gained attention and which did not. It was an entirely new architecture of
information that would have profound consequences. But at the time, it was
a new world we learned to use to our advantage without giving it a name.
We knew that we could tell an entirely new kind of story on our own and
reach people by leveraging new connections while still embracing traditional
media tools. Whether we meant to or not, we were building multi-edge
relationships in a graph (a concept further explored starting in chapter 2)
where the people were the most numerous and most powerful type of node
in aggregate. We understood that we weren't the center of attention—and
that relevance and attention were no longer going to be the same thing.

Leveraging this new architecture of information, President Obama's
victory was glorified by one half of the country reveling in a "post-racial"
America and lamented by the other as symbolic of the ever-accelerating
collapse of traditional American values at the hands of liberal coastal elites.
The idea that eight years of "Obamaland" gave way to "Trump's America"
is a ridiculous oversimplification; it reflects a liberal bias and an elitist
cultural understanding that are emblematic of our misunderstanding of
the world we live in. This tendency toward either-or simplification—a
common cognitive shorthand—is exacerbated by a media landscape aller-
gic to nuance. There never was an Obamaland independent of Trump's
America; both communities coexisted, however separate, and experienced
the 2008 campaign together. But in the end, they were left with completely different conclusions.

The United States is a large, diverse, complex country. Although 2008
may have felt like the ascension of intellectual coastal elites over everyone

else, that is only how it felt on the coasts. History may be written by the victors, but that doesn't mean it is written accurately.

**BENEATH THE SURFACE** of this new information architecture was the start of a cultural fraying that we now experience as a constant feature of American life. The public sphere, as defined by Jürgen Habermas in his 1962 book *The Structural Transformation of the Public Sphere*, represents both the physical and media spaces where people gather as a public to define the needs of society and refine the values that guide the state. Importantly, it is also the mechanism that legitimizes the state and its policies through connection to public debate. The American public sphere was beginning to collapse under the pressure of commercial mass media. The "declaration culture" of social media—the constant posture of declaring intent and truth without having to demonstrate or embody either one—was already pushing us back toward the representational culture of feudal dominance that we had left behind during the Age of Enlightenment. We were retreating from the post-Enlightenment culture of pluralism and reason, which had created the critical cultural spaces that made both the American and French revolutions possible.

Around the fringes of the 2008 campaign cycle was an extreme, angry reaction to the elitist, coastal, liberal ascendancy that President Obama represented to many people in the country. This reaction was often the expression of a more nuanced frustration, a sense of being left behind by an increasingly digital and increasingly unequal economy unconcerned about huge swaths of our country. But ultimately that anger would be categorized and largely dismissed by another oversimplification: latent racism in America.

Racism is both a foundational and a current feature of American culture and institutions. Rather than merely an awful historical truth, it still animates—both consciously and unconsciously—much of our institutional power and thinking. The tension and anger evident in the late stages of the 2008 campaign were also animated, however, by a growing disconnect between the realities of American life among disparate communities that

shared fewer and fewer connections and lived in greater and greater isolation. With no shared public sphere, we were starting to lose our capacity to
share experience and to debate. We were becoming two Americas talking
past each other about two different realities.

On November 9, 2016, I awoke to another bleary-eyed fall morning,
this time in New York City and on the other side of another profound
political upheaval. The same America that had elected Barack Obama eight
years earlier had just elected Donald Trump our forty-fifth president. The
2016 campaign cycle was not a competition over the core questions facing
America or even about competing answers to the same questions. It was the
first campaign cycle where we saw two entirely separate, parallel campaigns
operating in geographically coexisting but almost entirely disconnected
communities of voters—two campaigns competing over different voters in
different language about different things. The only questions in terms of
winning and distribution of power were just how big each conversation was
and how effectively motivated each community was to participate. The ability to debate had collapsed almost completely; the goal instead was for each
campaign simply to produce more voters. The trend, which had emerged
slowly over the course of a decade, became the dominant norm throughout
much of American culture and politics.

Donald Trump's campaign—itself an engine for exploiting dissatisfac-
tion, inflaming anxiety into anger, and directing that anger at other Americans, all in service of a mad grasp for power in spite of a general lack of
interest in the actual challenges of governing a large, diverse country—
was built on this division. The perceived unlikeliness of his 2016 victory
by nearly every establishment political operative (including me) and nearly
every mainstream journalist in America was predicated on the same invisible disconnect that began to emerge in 2008. The reality experienced by one
half of the country that would lead tens of millions to vote for President
Trump across two elections was right there in plain view, but without any
reference point for accurate interpretation by the other half.

It's hard to recognize the qualities of the air we breathe every day.
Coastal elites and Washington insiders recognize the world as a collection

of increasingly global disruptions but, like fish unaware of the water around them, see and feel very few consequences of those disruptions on a daily basis. While they see disruption as a business model or an intellectual exercise, many of our fellow citizens suffer the pain of that disruption every waking moment. This unseen, unheard pain has grown into a restless anxiety, often expressed as anger, that tinges their worldview every day—unmitigated apprehension about even the possibility of living lives of meaning and value, to provide for their families, to see their children reach beyond their own accomplishments. All the grand promises of the American Dream seem to apply to a narrower and narrower segment of Americans. Meanwhile, more and more people are deemed unessential to our country's future, in need of retraining and transition assistance—just an inevitable consequence of what economist Joseph Schumpeter dubbed the gale of creative destruction inherent to industrial progress.

While Black and brown America has lived with this (and much graver) anxiety for generations, working-class white America has just begun to feel this grating, growing panic for the first time over the last generation or so. According to a 2019 Pew study, while only half of white Americans remain generally optimistic about the next thirty years in America, more than two-thirds of Black and brown Americans feel the same—partly because they've always lived with these pressures, and partly because they are more likely to live in cities where vibrant economic conditions, creativity, and opportunity are more of the norm (even if difficult to access). This unevenness and uncertainty is entirely new to most of white America, who have gradually become markedly more pessimistic about the future over the last couple decades according to a *Washington Post* survey in 2015. Inflamed by this panic, the debate over class in America has commingled with our reckoning with race and turned into a shouting match.

In our new world, where we have greater access to increasingly isolated sets of information, this conflict is not playing out as a grand debate about how we move the country forward and define a future for everyone. We are not meeting in a public sphere to share familiarity of experience, to find commonalities amid our uncertainty. Instead, just like the 2016 presidential contest, our conflict is playing out in parallel, instantaneous conversations: one about global interconnection, growth, and infinite opportunity, and one about the collapse of American exceptionalism, a tenuous connection to the American Dream, and rising anger at being not only ignored but explicitly deemed irrelevant. Because they are parallel, these conversations don't intersect. Because they are instantaneous, they have no memory or shared context.

Increasingly, we live in two (or more) Americas. Our civic life reflects our complete lack of debate. Our two major political parties campaign on their conversation alone and wonder how the other side could possibly be so out of touch. The 2012 election may have been the last where two opposing sides competed over the core issues that should define American civic life.

In our new modern media landscape, as reflected by the complete divergence revealed by the 2016 campaign cycle, there is no persuasion. No debate. No meaningful argument. No productive conflict. No familiar diversity. No need to coexist with people with whom we disagree. But this isn't just the collapse of American politics in the face of a massively self-centered president.

President Trump was merely a symptom of a much larger shift in American culture, propelled by the rise and growth of a whole new information landscape that has completely altered how we consume information and understand the world around us. We feel lost and disconnected, desperate and isolated—not necessarily because we have actually grown so far apart from our neighbors, but because we misunderstand how the world now works. We fail to grasp that we have been intentionally cut off from each other for the benefit of the media systems and technology platforms that are ostensibly meant to connect us. In our heads, we carry an old map of the way information used to flow—a set of norms and expectations based on the past five centuries of media and communications. We refer to cultural stories about that world that may never have been exactly true (like the myth of unbiased journalism). And we feel disoriented because that map no longer accurately represents the reality in which we live.

The old model provided stable channels of consumption and forced many of us into shared experiences. Perhaps we wished for more options and a way around the narrow norms enforced by the old model's gatekeepers (read: white male hegemony), but at least our shared experiences created a useful overlap that helped maintain shared spaces. This often uncomfortable proximity forced us to engage in a constant and necessary debate. That debate—the heterodoxy that makes our ideas strong and the tempering that makes our institutions resilient—is what makes America. Yet as we've been forced into narrow, isolated communities without overlap, our ideas have become homogenous and weak. Our grand, necessary debate has given way to a self-reinforcing performance art of preaching to separate choirs in separate churches.

By holding on to the old model in our minds, we remain trapped in our view of a world broken, rather than renewed, by innovation. We are used to understanding authority as an institutional feature, but in a world where relevance is driven by attention, authority is easily conflated with celebrity. Popularity masquerades as credibility, and credibility seems like a function of little more than repetition and dosage. Instant gratification is painstakingly designed into the interfaces we use to consume information. And from our uncertain vantage points, amid the comfortable haze of confirmation bias and dopamine addiction, we begin to substitute proximity for everything. This flawed understanding can be easily gamed and algorithmically manipulated to benefit whoever controls the algorithm for whatever goals they might wish to pursue.

In their desperate attempts to maintain relevance on the same outdated, misshapen map, news companies tighten their embrace of attention-seeking content while holding on to the pleasant fiction that they are optimizing for truth. In actuality, they are only accidentally slipping into truth when it coincides with their optimization for attention.

THE COMFORTABLE, CHANNEL-BASED view of paid, earned, and owned media, with media companies transmitting information from publishers to audiences, trapped American consumers as stable, passive audiences in a hierarchy of information. But that is no longer where we live.

Today, we live in a massive graph where we all—media companies, publishers, individuals—consume, create, and distribute information. For each of us in our various places within the graph, the view is unique. Up to now, we have failed to recognize how much more power we have in this new world and how much more responsibility comes with it. And that is precisely why we are disoriented.

The media and technology landscape in which we live does not determine or control all of culture—but if culture is our shared stories, then how we tell and share those stories is fundamental to who we are and how our society functions. Investigating how media and technology have evolved to create the America in which we find ourselves, without our permission and in ways that we understand poorly, is essential to finding a new, better path forward. Healthy, vibrant, pluralistic democracy is not incompatible with modern media, but our failure to understand modern media weakens the foundation of American democracy. It makes us vulnerable to commercial and political actors whose incentives do not necessarily align with our public good. By allowing faithless actors to define the core principles of this landscape, we have abdicated our own public authority when it comes to laying the foundation for a new, modern public sphere that can deliver on the promises of greater interconnectivity rather than separate and exploit us for profit.

But we can take it back.

If we can make these implicit principles explicit requirements of redesigning platforms and restoring institutions, then we—*the people*—can reclaim the power we hold as the most numerous type of node in the graph. If we are prepared to lead confidently in the direction of a moral, public sphere meant to challenge and uplift all of us . . . then we can still have the jetpacks.

# 1

## LOST LIKE A GOAT IN A PHARMACY

I N HIGH SCHOOL, I had a teacher who taught both Spanish and economics. Señor Muñoz was my first guide into the world of macroeconomics, into the interconnected systems of markets and effects that made up the commercial world. Baffled by the tension between rational expectations and irrational behaviors, I drove Señor Muñoz batty with my incessant questioning. He would often drop his head and declare, with a heavy sigh, "Mike, you're as lost as a goat in a pharmacy."

That image of a goat—calm, unbothered, lost in the useless aisles of a pharmacy—has stayed with me ever since and often comes to mind when I'm confronted with systems that are at odds with the people for whom they were designed.

Our media and information ecology is the pharmacy.

And we're the goats.

Just as there is nothing wrong with the goat—there is nothing wrong with us. The dislocation we feel and the dysfunction we witness are not the inevitable consequences of modernity, of technological progress, or even of "those ridiculous millennials." And all this confusion is definitely not our fault. These feelings are instead the inevitable consequences of the dramatic expansion of media systems that have given some individuals more power, but none of us more clarity—systems that offer more information than ever yet obscure our choices. These systems promise greater connectivity and joy, but were designed to exploit our attention in exchange for money.

Our modern media systems provide the architecture for us to access the stories we rely on to make sense of the world. And these stories—to crib Yuval Harari from his 2011 book *Sapiens*—are the scaffold of history and culture. Storytelling is, in many ways, humanity's superpower relative to the rest of the natural world. The systems we rely on to publish, distribute, discover, and consume stories have expanded that capacity exponentially. When our ability to tell stories expands, as it did with the invention of the printing press, the radio, the television, and now the internet, we expect the world to feel smaller. We expect to understand one another in new ways and to experience greater empathy. We were promised a cyberutopian vision of a better civilization: greater connectivity inherently leading to strengthened Western liberal traditions and greater social cohesion. While this expansion of our storytelling ability may have been initiated with the public benefit in mind, there was no public declaration of intent, no expression of values and beliefs to ensure that it would improve our public sphere or grow our civic life.

Robert Kahn and Vint Cerf started defining the TCP/IP protocols that still govern much of the transmission of data online at DARPA in the late 1970s. Cerf left DARPA in 1982 to join MCI (now Verizon) and begin commercializing the internet. It was another full decade later, in 1992, when they created the Internet Society as a home for the ongoing technical standards work necessary to support the internet. And it wasn't until the summer of 1999—almost a decade after Tim Berners-Lee published the first website, introducing the idea of the World Wide Web—that Cerf and Kahn realized the Internet Society needed a "global brand" as a tool to advocate for the kind of open, available internet they had always envisioned. For more than two decades, the commercialization of the internet was unguided by anything other than entrepreneurial drive and the implicit values built into its protocols—and the public benefit has been trying to catch up ever since.

The Internet Society was established as an advocacy platform (not just a technical one) one year *after* the 1998 launch of Google—where Cerf has worked since 2005. And unless you work in internet technology, you likely have never heard of the Internet Society, which says something about the reach of that public advocacy. Other efforts similarly failed to reach the kind of cultural resonance necessary to meaningfully shape the development of modern media. From the search battles and browser wars of the first dotcom boom in the 1990s to the rise of social media—beginning with Friendster in 2002, Myspace in 2003, and then Facebook in 2004—the rearchitecting of information by social media was driven by venture-backed companies, not public advocacy groups.

At this new frontier of public information, private businesses with no incentives beyond rapid, venture-scale returns operate in the absence of shared moral, legal, or philosophical guidelines. The problem with the move-fast-and-break-things, disruption-is-good culture is that these companies end up breaking things indiscriminately.

To generate those kinds of returns on that kind of timeline, companies like Google and Facebook become laser-focused on the certain user behaviors that drive growth. They design their systems to promote those behaviors, without seeing the unintended consequences of the experiences they have created. In the design process, commercialization crowds out other values; hypergrowth and the chase for rapid scale eclipse ethical questions. Companies like Facebook and Twitter have replaced casinos as the greatest behavioral psychology experts of our time. They promise us connection, togetherness, and convenience in exchange for exploiting our attention to fuel their growth. We are left confused, overwhelmed, and disconnected, munching on stale Halloween candy in the seasonal aisle.

It turns out humans are largely as predictable as goats.

From an evolutionary standpoint, we are well trained by a few hundred thousand years of survival to pay attention to certain kinds of inputs. In his book *The Organized Mind*, Daniel Levitin goes to exhaustive lengths to break down how our attention works, what our mind has evolved to handle, and how we react in a world of information overload. We instinctively avoid existential threats—saber-toothed tiger! speeding city bus!—yet discount other risks based on optimism bias: our tendency to underestimate the likelihood of negative events. This bias is what helped our ancient ancestors leave the cave in the first place, and now it lets us drive our kids to school without flinching at every bus that speeds by. These well-ingrained reaction paths trigger an enormous amount of our automatic behavior and subconscious decision making. And our predictability makes us easy to manipulate.

If I can obscure the potential likelihood of something negative happening, you will largely ignore it in order to avoid paralysis.

On the flip side, if I can scare you, I can keep your attention because fear indicates something that might affect your potential survival.

If in response to that fear, I can also offer you an immediate mechanism to push back, to express your outrage at the offense—to go from being pushed back onto your heels to leaning forward onto your toes—you will jump at the comfort of realigning your posture. And in a world of constant perceived attack, this outraged reaction becomes our default.

Media companies and technology platforms know our psychology and these behavior patterns inside and out. They design algorithms that present us content and offer us opportunities to respond to that content that ensure we stay engaged. Facebook needs you to stay on Facebook as long as possible to maintain its inventory of attention. If it needs to maximize your attention, how likely is it to optimize your experience for anything else? It will show you content that scares or outrages you, to keep you engaging in order to assuage those threats. And it will do so in a way that obscures the negative consequences.

This may not be a conscious effort to promote content that scares and undermines, but so long as sustained engagement and attention are Facebook's primary goals, our psychological vulnerabilities and automatic responses will push the company toward peddling outrage.

Every. Single. Time.

Our collective sense of dislocation and uncertainty about where we are headed is not an inherent feature of modern American culture. It is a feature of a media and information landscape that was designed to maximize the conversion of attention (often negative attention) into money. A public sphere built around fear and outrage is not the kind of commons required for a healthy civic life. It is the product of a business model meant to exploit the potential for hyperpersonalization implicit in this new land-scape—all for the benefit of companies that claim they want to connect us. When in fact, these companies reflect an extractive mentality that appears uncannily similar to the industrial-era companies they seek to disrupt and displace.

The dislocation and uncertainty we feel and the unhealthy public sphere we inhabit are driven by the consequences of that commercialization and exploitation. This dysfunctional model leaves us dependent for the com-munication of story, identity, and culture on systems that are purposefully designed to outrage us. We lack the shared stories that we need to remain connected in a single, albeit diverse society. We have no room to find com-mon ground with difference, no real opportunity for discourse, no chance to learn. The habit of outrage results in a "declaration culture" dominated by the constant declaration of position and status rather than the demonstration or embodiment of belief. As a result, we are driven to engage with each other based on stereotypes and caricatures and left largely unable to debate safely.

The public sphere we rely on for society and civic life to function is being actively eroded by the very systems of media and information that we rely on to connect us and bring us closer.

## WHOSE AMERICAN DREAM?

America is large, diverse, and complicated. The idea that we ever held a consistent, shared story of what it means to be American contradicts the obvious reality that in every era of American history, significant segments of our population have been excluded from the story. Indigenous peoples, women, people of color, immigrants have each experienced "the American story" in a profoundly different manner from the experience of those in power in each era. Our founders described equality and freedom in elevated and ambitious terms, yet excluded non-landed men and all women from the expression of citizenship, and institutionalized slavery and the partial value of enslaved people.

Among those with access to the political rights of citizenship, the story of an emergent America—independent from an anachronistic monarchy an ocean away—was consistently held by those founders. How to form a "more perfect union" was a matter of profound disagreement, but in service of their mission of claiming national self-determination in an entirely new way, they held a shared foundation story of this independent, emergent America.

That founding story slowly evolved into the familiar twentieth-century story when, in 1931, historian James Truslow Adams wrote his book *The Epic of America*, canonizing the story of America as "the American dream, that dream of a land in which life should be better and richer and fuller for every man, with opportunity for each according to his ability or achievement." This self-reliant, optimistic account of the United States as a country of opportunity was written during the Great Depression as a confidence-restoring narrative for a country suffering. And even then, we continued to exclude whole communities while attempting to maintain a uniform, unifying idea for our foundation. Only particular types of ability and achievement, which were synonymous with whiteness, were rewarded as the American Dream.

Different versions of that "all-American" story were confined to narrow audiences—would-be revolutionary conversations that would slowly find their way into the mainstream and force changes in who could expect to be part of that broader, mainstream narrative. The women's suffrage movement of the early 1900s and the civil rights movement of the 1950s and 1960s expanded who might expect to find themselves included in the foundation story, but it remained largely intact as a dominant feature of our national culture.

As the Depression era gave way to World War II and the "boom" era (the start of what *Time* publisher Henry Luce dubbed "the American century"), followed by the accelerating inequality of post-labor America in the 1980s, our understanding and expression of that founding story began to crack and shift. It took on an increasingly selfish and commercial tone. It began to fragment along with our media and information landscape. As our sources of information multiplied, new voices with alternative conceptions of who we are as a nation found their way from the revolutionary margins to eager audiences and, eventually, into the mainstream of American culture. This slow breakdown of the large, stable audience led to the breakdown of the large, stable narrative. The ever-present failure of that story to include every citizen was revealed and unpacked in greater and greater proportions. The story of the American Dream began to fail as a foundation for national culture.

Post-9/11 America craved a coherent national identity just as America did in 1931, when the American Dream was first characterized as a unifying idea. Yet in the run-up to the second Iraq War, the firm stratification of news audiences and the fragmented idea of America played out as a textbook example of how broken our national narrative had become. The events of 9/11 and the immediate retaliation in Afghanistan provided a final blow to this cracking model of American identity.

The single national identity was a myth.

A large audience of mostly white, increasingly rural, generally more conservative, and mostly Republican Americans were presented a story of an aggrieved and wounded America ready to retaliate against a violent world with the audacity to challenge American hegemony.

A more diverse, increasingly urban, more liberal, and mostly Democratic audience was presented a different story, about an arrogant America forced to wake up to the hidden costs of globalization, the pervasive lack of opportunity in the face of American opulence, and the enduring scars of colonization around the world.

According to Pew Research, which has followed polarization in American politics since 1994, the media habits of these two factions have diverged sharply since the dawn of digital media. Back in 1990, the march toward the first Iraq War was generally seen as certain and righteous; the alternative frames were seen as opposition to American identity; and the nation rallied around a "universal" patriotic cause. The divergent experiences of the second Iraq War would not have been possible in the old, channel-based, pre-internet media landscape.

Conservative audiences remain significantly more homogenous in both demographics and information consumption habits. In 2014, self-described conservatives consumed more than twice as much news from Fox News than from any other source. Liberal audiences, on the other hand, are more diverse in both identity and information consumption. Self-described liberals get news in more balanced proportions from CNN, major networks, and MSNBC, and more than a quarter also regularly get news from Fox News.

The consequences of this stratification in just a single decade are startling.

The narrative being consumed by the largest unified block of Americans is *always* the core conservative narrative exemplified by Fox News. Meanwhile, dozens of related (but not uniform) liberal narratives find community in smaller, often competing groups. So when mainstream media outlets begin to normalize the various narratives into a national story, the conservative story generally sets the frame for conversation.

The self-reliant, semi-libertarian myth of the 1930s American Dream remains a fundamental expression of an almost nostalgic brand of nationalism in largely white, largely conservative communities. Yet by the time the internet had emerged as the foremost source for media and information, a 2015 survey by the Harvard Institute of Politics revealed that half of 18–29-year-olds flatly believed the American Dream was dead for them. Suffering both a loss of ubiquity and a collapse of belief, the narrative that once unified us is now translated differently, reframed, and expressed in new ways, all depending on the stories we consume and where we get them.

As the internet grew, the opportunity to tell new stories spread, and the shared story of America fractured at an accelerating rate. Yet the homogeneity of information in conservative communities continues to isolate them from media content that reflects communities where the American Dream is not a triumph, but a cruel and mocking reminder of what is not possible, of what is withheld or taken. White America has never, until now, heard alternate American stories at significant scale. But now that these excluded communities have more equal access to storytelling power—the capacity to tell *their* version of America—instead of the mere whisper of an oppressed minority, it has become a competing understanding of the core narrative of America.

There are still profound biases at work in corporate media, but the base architecture of how we share information has made those biases less definitive. New narratives have escaped from alternative subcultures and moved into the mainstream fight over our national identity. Only a few years ago the idea of reparations for slavery was considered a fringe political theory confined to discussion in the history and economics departments of HBCUs (historically black colleges and universities); in the 2020 Democratic presidential primary, more than half of the candidates made it a proactive pillar of their platform.

For the first time, we have more than one national identity story competing for *mainstream* dominance. But that competition is happening in a media landscape that encourages us to primarily consume whichever version of the story we already accept.

A true national story not only gives us a useful narrative to lean on in times of crisis, but also provides a shared lens for interpreting new information and events. A shared core story creates the basic framing for how a citizenry interprets news, events, analysis, policy, leadership, and even cultural reference. A splintered understanding, on the other hand, reinforces the long-standing disparity of experience that has created an era when the exact same facts are carelessly interpreted in fundamentally different ways. And as individuals from oppressed communities make their way, against systemic odds, into the ranks of the powerful and access the rights of citizenship, that alternative story increasingly becomes part of mainstream American culture.

The problem here is not diversity or the inclusion of more voices, but an inability to embrace that diversity together in a unified experience. Without measured discussion and debate, we are unable to reconcile that diversity of experience into a shared identity of a diverse *and* unified nation. And when we lose our shared story (however flawed), we cease to be members of a shared community. We have conflated being part of a shared narrative with believing the same things and agreeing all of the time, and we've given up being in community with people with whom we disagree. Instead we become citizens of multiple Americas that share only geography.

## WHERE DID WHO GO?

Multiple stories of a national self aren't necessarily at odds with social cohesion or effective self-government. In fact, vibrant debate, even (or perhaps especially) about the nature of the community, is essential to the effective functioning of a democracy. But multiple stories of self that never intersect—when two citizens of the same nation don't experience each other even fleetingly and don't even recognize each other as fellow citizens in a shared polity—are troubling at best. At worst, this state of affairs is disruptive if not outright dangerous. It is entirely possible to make space in the national narrative for our diversity and our parallel histories, but we aren't attempting to do that. What's more, the media and information systems that support our storytelling and information gathering are actively working against that possibility.

Modern communication and storytelling have given us access to a greater diversity of voices than this country has ever known. The internet promised to make the world smaller: to give us visibility and insight into communities close and far that are different from our own; to offer new perspectives; to expand our capacity for empathy; and to make us more open to shared citizenship with people who hold different opinions. But we don't share a common public sphere where we can openly engage and debate that shared story. The utopian view of greater connectivity leading to greater empathy is predicated on behaviors that platforms like Facebook do not encourage. On the contrary, the existing platforms undermine that unity at every opportunity by their business models and their design.

Eli Pariser, in his 2011 book *The Filter Bubble*, details the lengths to which the powerful platforms of our modern public sphere are forcing us into smaller and smaller circles, where we hear constant echoes of our current thinking but experience less and less of the expansive diversity all around us. We are losing the sense that the world around us is complex and full of variety. Our typical experience of society is more insular; our nearest, most frequent connections all think alike and consume relatively similar information sets. We eagerly band together, certain that we are surrounded by dangerous "other" groups of people who oppose us.

Part of this turning inward is a reaction to complexity and uncertainty. Humans generally aren't well suited to handling the kind of complexity and breadth of choices available to us in a globalized, interconnected world. Levitin's *The Organized Mind* identifies that on a day-to-day, minute-by-minute basis, we live in a constant state of information overload that is wildly beyond our conscious computational threshold. We crave ways to summarize, to order, and to limit the scope of that overload so we feel less overwhelmed. Exploiting that need for simplicity is a set of tools whose business models are dependent on experiences that optimize (read: control) our behavior narrowly. The combination of these factors drives what Pariser dubbed "filter bubbles": the small, self-reflexive realities that we live in. Inertia alone means we have to actively work to break out of our bubbles. Add in our easily exploitable psychology, manipulated through the design efforts of brilliant, multibillion-dollar industries that reinforce our narrow experiences, and breaking out is getting harder and harder.

A healthy democracy relies on the expression and exploration of a broad range of ideas, and the spectrum of ideas that are politically acceptable to society is known as the Overton Window. When we experience the world through an ever-narrowing aperture, the Overton Window begins to collapse. What remains are smaller, discrete social segments that fail to overlap and end up excluding everything with which they disagree. This narrowing of our individual perspectives, together with the breakdown of a shared gradient of thinking, eliminates the concept that some ideas are unacceptable. We see each other only in the ways chosen by ourselves or people like us. We are told about "others" generally through a lens of threat or conflict. We rarely see those others in their own words, and we rarely express ourselves to them directly. This failure to seek a genuine perspective on others, combined with the collapsing window of acceptable ideas, has become a core feature of American culture and politics. When all "others" are threats, all disagreement is existential. This encourages mainstream extremism, where there is little effort or incentive to understand anyone who does not already belong to one's worldview.

In a world steered by confirmation bias and urged toward homogeneity, we lack a shared public sphere where we overlap with other communities. There is almost no place we all experience together—nowhere to meet on common ground, nowhere to experience each other directly, and nowhere to see each other as whole and complex humans. In today's America, liberals and conservatives are told about each other in one-dimensional caricatures that reveal little of our shared humanity. This limited perspective makes us easy to villainize; it reinforces our differences and the threat of "other" in support of our current worldview. Combined with rapidly expanding economic inequality, it creates the ideal conditions for cynical political machines to pit us against each other in angry populist or nationalist conflict.

In this constant disconnection from the whole, our citizenry also lacks historical context—even along the shortest timescales. We live in constant, instantaneous streams of information completely divorced from what came before or what will come after, which makes it nearly impossible to remember what was said or what was promised, much less whom to hold accountable. Without a reliable memory, without the ability to build authority over time, without clarity of authorship and ownership, we cannot build trust or reliability. Instead we remain bound by the instantaneous choices platforms optimize for us in their desire to keep us reading and clicking, with no meaningful context.

The sense of timelessness makes the outrage all around us feel constantly fresh, unlike the stale artifacts of grudge and anachronism. The threats we sense feel existential rather than common and toothless. Disconnection in time also makes it easy for storytellers of all kinds—but especially our leaders—to make promises no one will hold them to and to hide from previous statements under the tyranny of the moment. Our memory grows shorter while the technology platforms we rely on capture our history as inputs to their algorithms. They leverage what they know about us not to create better understanding among us or to develop more comprehensive context for our conversations over time but to optimize our momentary choices for their financial gain.

## WHOSE LINE IS IT ANYWAY?

For democracy to work, we have to be able to argue productively. The ability, the need, and the responsibility to debate our beliefs—about the kind of country we want, the kind of leadership we want, and ultimately what rules should govern society—are fundamental to self-government and to good, active citizenship. Debating is also how our ideas get stronger and how the quality of our thinking improves over time. It's how our perspectives widen and our empathy becomes more inclusive. But down-stream from an America with more and more stories of self is a country where choosing a particular story of self has become an existential expression of identity. And when something as intrinsic as our identity is questioned, we react as if confronted with an existential threat.

The more our civic conversation is framed in existential terms, as "us versus them," the less we are able to debate productively.

We already lack a shared foundation for our national identity and con-versation. Now we are at risk of compounding that with the suggestion that *me being right* and *you being wrong* means you have no right to participate—or in some cases, to exist at all. This kind of exclusion dramatically under-mines our ability to discuss what we want for ourselves and the country.

Existential conflict is an extremely effective frame for motivating groups to seek the acquisition of power over one another, and it has become a central feature of electoral campaigns. Narratives like *If we don't win, we die!* and *This is the most important election of your lifetime!* drive voter participation because everyone is incentivized by their own survival. But they also drive an orien-tation toward difference and conflict and an expectation that disagreement is equivalent to existential threat. They create the pretense of an inexorable, unsolvable reality. And they introduce language of violence (e.g., targets rather than communities or audiences; enemies rather than opponents; kill rather than defeat) to everyday political conversation in a way that rein-forces our threat response and makes disagreement potentially dangerous. Leaning toward conflict ultimately makes us less likely to seek out different perspectives, to debate and argue in safety—because it doesn't feel safe.

How many of us are encouraged to avoid talk of politics at Thanksgiv-ing dinner? On dates? In the workplace? How many of us avoid politics entirely? Nothing could be unhealthier than a democratic polity unwilling to debate, but increasingly that is exactly our experience of politics: either ardent activism or active avoidance.

When our stories of self diverge and our differences are coded as exis-tential, it becomes more difficult to recognize our shared experience. Take, for example, the collapse of opportunity and the rise of deaths of despair in former manufacturing towns of the Midwest and the profoundly similar experience in inner cities during the 1980s and 1990s. The underlying causes are wildly different: one breakdown driven by modernization and globalization, and the other by systemic racism. Even so, communities hollowed out by forces beyond their control and left to criminalize their own survival ought to be able to recognize (and potentially organize around) shared pain and experience. But the history of those two communities is told in very different narratives: one about health ("the opioid crisis") and one about violence ("the war on drugs and gangs"). When those stories are interpreted through entirely separate lenses—an economy stolen by immigrants, public systems propping up urban welfare queens, the earned pain of oppressors getting what they deserve—the potential for shared experience evaporates.

Turning debate into existential identity warfare is dangerous. It makes compromise undesirable, even threatening. In a two-party system of representative self-government, the inability to compromise is akin to an inability to govern. We may take turns imposing our political will on the other side in a seesaw death match of ideological entrenchment, but we never make real progress. Our ideas never get stronger. As a nation, we stagnate.

## REPRESENT WHO?

Leaders who want to hold on to power speak to those they believe are listening and willing to act. The more our leaders interpret "citizenry" as "activists," the less they will even attempt to consider nuance or compromise. Leadership in the United States, representing the wide range of opinion and perspectives, once produced voting patterns in Congress with significant variation. Now, like votes with like.

Looking back over Pew's political polarization study that started in 1994, overlapping perspectives used to be the norm. Twenty-five years ago, 64 percent of Republicans were to the right of the median Democrat, and 70 percent of Democrats were to the left of the median Republican. In 2014, those numbers were 92 percent and 94 percent, respectively.

According to a parallel study led by Clio Andris at Penn State, the likelihood that any two members of the House of Representatives from different parties would significantly agree on positions across party lines went from a high of 13 percent in the 1970s to just above zero in 2009. Only in those rare districts that exist along ideological and demographic boundaries do we find leaders interested in compromise. In a government that has come to mirror the divergent stories of self and existential us-versus-them culture of the country, these more bipartisan leaders end up being the most vulnerable—and therefore the shortest serving.

Today, we are left expecting nothing like real leadership from our leaders—and we are getting largely what we expect. Lack of participation is a self-fulfilling spiral. Reversing that spiral is at the heart of what is required to reclaim our civic life.

Our political reality has descended into the hyperpartisan swamp we claim we want to avoid. Our hyperpolarized ideological climate rewards politicians who demonstrate extreme thinking and positions. A partisan process—gerrymandering—determines representation and civic boundaries. Our campaign finance system permits almost unlimited, unregulated spending in political campaigns. Public conversation and participation, underpinned by nearly absolute polarization of perspectives, leads to partisan leadership—where compromise is a bug, not a feature.

This is how governing begins to fail.

Our elected officials, ostensibly bound to represent all of their constituents (not just those who voted for them), respond to ever-narrowing groups of donors who fund them, lobbyists who are paid to influence them, and activists who voted for them in primaries. These small subsets of citizens tend to represent either of the edges of American ideology on both sides (extreme liberal and extreme conservative) or single-minded corporate interests. This shift toward extremism and corruption leaves a large swath of Americans, up to 70 percent, as an "Exhausted Majority"—as identified and dubbed by the Hidden Tribes project of More In Common—who feel ignored by a system of extremism. More and more of us want less and less to do with politics. And so we disengage from our civic life.

When we disengage en masse from a system of representative self-government, we cede power either to those only interested in personal power or to those who want to use public systems for private benefit. Neither category is interested in our collective success.

Voting is an imperfect but convenient metric of civic engagement. It represents, to some extent, the baseline—the least engaged we ought to be as citizens. Since the Nineteenth Amendment institutionalizing women's suffrage was passed in 1920, presidential election turnout (calculated by votes as a percentage of voting-age population) hasn't varied much from an anemic percentage in the high-50s to low-60s. Over the past half century, however, voting in city elections has plummeted. Los Angeles and New York elected mayors in 1950 and 1953 with 70 percent and 90 percent turnout, respectively; in 2013, those numbers were 23 percent and 26 percent. In smaller municipalities, these numbers get even worse. According to a Portland State project in 2016, about 15 percent of eligible voters nationwide vote in mayoral and city council races.

Politicians look at these trends and decry the painful apathy of the American voter. Although, many other factors also affect turnout—from lack of competition to ballot access issues that disproportionally affect younger voters and underrepresented communities, this disengagement is largely rooted in the increasing belief that voting and government are either too corrupt to participate in or too useless to matter.

As a nation, we see political leadership shifting away from a culture of public service toward careerism and private benefit, and we interpret that shift as corruption—and rightly so. Yet when our skepticism leads us to disengage, our government becomes captive to these narrow, often economic interests. The end result is that our trust in government plummets, and the cycle begins anew.

The Edelman Trust Barometer research project has quantified over the past two decades a collapse in trust of key public institutions, after a downhill slide that began in 1896 and was dramatically accelerated by Watergate. Edelman scores institutions on a multi-axis trust framework that includes both ethics and competence scales. Government is near the bottom of both scales, with media only slightly better off. No institutions on the framework are considered both ethical and competent. And Americans now overwhelmingly trust "people like me" more than they trust public institutions or even subject matter experts.

Embracing community opinion isn't in and of itself problematic, but that's not what is happening. Our embrace of "people like me" is a function of the flight from institutional leadership. We now substitute proximity for authority. When you combine expectations of corruption with lack of trust and the sense that institutions don't have a vision for the future, it begins to make sense why most Americans disengage from politics.

Filling the void left by our unwillingness to engage is a system dominated by our ideological extremes, where we are convinced that compromise is failure and are uninterested in debate. Only the loudest voices are heard at all—mainly activists preaching to their choirs. Activism is a valuable, powerful, and essential form of civic engagement, but it should not crowd out and silence less ardent citizens who are just as deserving of opportunities to use their voices, however quiet. Rather than reaching farther and listening harder for these voices, however, our leaders move toward narrower constituencies, and the general public pulls back even further from political engagement. Our feelings of isolation and disconnection increase, until we have the sense that we are not represented at all.

And the cycle continues.

## WHO LEADS?

Our democracy is a system of self-government that was never supposed to have a ruling political class, and yet today we rarely know the people who represent us. We've ceded the swamp to the swamp monsters.

Back in the 1980s, House Speaker Tip O'Neill popularized one of modern politics' most enduring political aphorisms when he told America that "all politics is local." At that time, *local* was still a pretty good proxy for what mattered most to most people, as local issues still dominated the everyday experience for most Americans. But what he really meant was that all politics is *personal*—that what matters most to *me* is unique to *me*. And the more we recognize each other's uniqueness of perspective, the more willing we are to take that perspective into account. So the more community and experience we share with our leaders, the more likely they are to effectively represent us. But the systems of modern media and information are working against that kind of symbiotic civic operation.

Knowing our leaders may seem like an odd feature of privilege, but that feeling itself represents the problem. At the same time average Americans have turned away from politics as something useful and meaningful in their lives, campaigning has become a multibillion-dollar industry. Stumping for most higher offices demands a full-time commitment that requires most candidates to give up their day jobs. Not only is running for office perceived as far less useful and more self-serving than it once was, it is also less of a viable path to service for people who are truly civic minded but cannot afford to run.

The inaccessibility of leadership is consistent with the inequality we see in our economy more broadly. Creating insurmountable economic barriers to running for office results in leadership that is not only disconnected from but often indifferent to the pain and experience of most Americans. Winners of an unequal economy are not well positioned to recognize the inequality of the system that benefits them. Too often, our elected leaders remain blind to social inequity in the form of systemic racism and sexism. They are unresponsive to an unequal economic environment where the real value of wages is disintegrating even as wealth among the wealthy expands faster than ever.

When leaders from underrepresented and vulnerable groups are present, it ensures greater awareness of those most in need of public and social programs and that those communities can exercise representative power. Lack of representation, on the other hand, perpetuates inequality and promotes the idea that certain groups are struggling because of their own failings rather than a failure of leadership.

When we lose faith that our leaders represent us, we lose trust in the institutions they lead. This distrust reinforces our sense that we are alone, forgotten—left to determine our own fate. This profound sense of abandonment, of isolation, of genuine loneliness makes us even more susceptible to the comfort of confirmation bias and the filter bubbles offered by our media systems. And our disconnection from each other accelerates.

## HOW MANY AMERICAS?

In 2004, before his ignominious fall from political life, former Senator John Edwards spoke at the Democratic National Convention, hoping to surface a similar sense of disconnection in America. In what is often referred to as the "Two Americas" speech, he railed against inequality—how two different groups experienced the American Dream in entirely separate ways. He recognized at that moment the possibility to confront this inequality, to see it clearly, and to reimagine a genuinely equitable, unified America—one with diverse opinions but a shared core identity. Yet nearly two decades later, the bifurcation of America has only accelerated.

The core economic inequalities of modern America represent only one lens through which to perceive that our country is indeed divided. As our media and information systems fragmented and were then rearchitected by the rise of social media into the new reality of the graph—a reality defined by unique perspectives, filter bubbles, and a dramatic decrease in shared experiences—we have introduced significantly more cultural isolation into American life. Where John Edwards saw America dividing primarily along class lines, today we could readily aggregate communities based on narrative lines. What we would see is in fact shared geography and little else—not two Americas, but 331 million Americas, with each of us seeing and engaging the world uniquely from our position in society.

Debate based on shared underlying principles is a foundational feature of American democracy. Without it, we risk losing the valuable, constructive sense of self-government. We miss out on collective striving for the best ideas and answers based on shared values, and instead we begin to imagine a monochromatic political society where everyone agrees and dissent is silenced.

Some elements to the dis-ease of American political life stretch far beyond the scope of a discussion of how information moves and how media functions. But this landscape has had—and is having—a profound impact on how we interact with each other, how we tell stories that define culture and norms, and how those norms and stories will be shared going forward. Culture is the sum of our shared stories. Without them, we don't live in one country.

As it turns out, our new media systems *could* offer greater diversity and *could* engender greater empathy. But inclusion is harder to monetize than outrage, and so the deluge of information separates us. Our choices collapse. Our worldview gets smaller. We feel adrift in a sea of threats rather than buoyed by new and broad perspectives amid a world of possibility. Disconnected in time by systems optimized for the instantaneous—the next view, the next click, the next story—we are left with no real memory, and no understanding of how to reclaim a shared civic life.

In our current scenario, creating a commons is nearly impossible. We need a new discussion about sharing uncomfortable proximity—about maintaining, and protecting as sacred and essential, a public sphere that we all share.

# 2

# I WAS TOLD THERE WOULD BE NO MATH ON THIS EXAM

**F THE DYSFUNCTION** that we feel every day is not our fault and is not inevitable, then how did we get here? We used to be great at storytelling. What happened?

The story about the evolution of media from Johannes Gutenberg to Mark Zuckerberg is only part of the story. Underlying this centuries-long tale of shifting formats and emerging platforms is a slow evolution of the architecture for how we tell stories, how we distribute them, how we find them, and how we consume them. And as a society, we have failed to keep up with that evolution. Our failure to understand and adjust to this new architecture has undermined both our ability to communicate effectively and the ability of our media systems to support the vibrant civic life we need.

"Media systems" are the complex, interconnected systems of people, content, technologies, and institutions that make up how humans create, distribute, discover, and consume information. The progression from oral traditions to the printing press all the way up through the rise of television has been one of constant expansion: adding and expanding the channels available to us for publishing and consuming ideas and information. Over time, although some formats have been replaced or lost altogether, media systems on the whole have become more complicated and have offered more choices. But in each era, the base architecture of information—a publisher creates, a media channel distributes, and an audience consumes—has remained largely stable. The basic idea of an information landscape consisting of various channels, whether they are books or TV stations, has been the modern reality of storytelling since the printing press was invented in the fifteenth century.

In the twentieth century, this explosion of channels collided with the escalating commercialization that had begun five hundred years earlier with mass printing. Forces of history and religion steered the impact of early mass media and helped codify culture; modern media, meanwhile, is dominated by massive multibillion-dollar corporations. Those companies have transformed one of our most fundamentally human and fundamentally necessary forms of identity creation—of communication, of culture crafting, and of history telling—into an advertising-driven profit machine. In turn, this accelerated commercialization has dramatically hastened the proliferation of available forms, networks, and channels while also introducing corporate-controlled mechanisms into the framework of storytelling that shift responsibility and opportunity drastically. This new information architecture has also transferred power and control from publishers and storytellers to platforms of distribution and discovery.

Through the 1990s and into the 2000s, at the pinnacle of mass media commercialization, we witnessed a major fragmentation. As cable television channels proliferated, large, stable broadcast audiences splintered into smaller, niche audiences, leaving companies struggling for competitive advantage and new opportunities to dominate access to increasingly specific communities. But while the complexity of reaching the masses continued to increase, the basic architecture of information remained the same. The power to create remained a function of professional storytellers. Audiences were harder to capture, but the same underlying structure remained intact: a publisher distributed content via a media channel to consumers.

As the 2000s wore on, something new emerged: with the rise of prosumer desktop publishing, more people were able to produce professional-quality storytelling. With the explosion of social media, the primary mechanisms of distribution were within reach as well. Thanks to this wave of innovation, the power to create became more accessible and, although still unequal, the power to distribute was now available to anyone willing to trade privacy for access to platforms like Twitter and Facebook. These new platforms enabled easy sharing and redistribution of content along new paths, knitting back together the channels that had fragmented and the smaller, disconnected, niche audiences that had splintered over the previous decade. Content that used to remain within the channel in which it was published was now free to travel easily and unpredictably to new, sometimes unintended channels and audiences.

This new architecture features characteristics that diverge radically from the centuries-old systems we're accustomed to. The consequences of these new characteristics—and our misunderstanding of them—are what drive our current national dysfunction and the degeneration of our civic life.

Gone are the days of stable, fixed relationships between publisher, distributor, and audience. Today, we consume information in a complex network where content is shared and consumed in unintended and unpredictable ways. In our old model, content generally remained in its published channel: a newspaper article was a newspaper article, and that was that. Gatekeeper power was centralized among the publishers, and the core choice was *where* to publish a certain story in order to reach a certain audience. In today's media systems, the roles are no longer fixed. Each of us is publisher, distributor, and consumer of information. Anyone or any organization can (and does) play any of these roles. Anyone can tweet just as readily as the president of the United States. Anyone can share a *Washington Post* article on any number of platforms and with people the *Post* never expected to reach. This new era has led to a paradox: despite having more access to more information than ever, we feel more disconnected and uninformed than ever. This contradiction is the defining feature of the new architecture and exploring it is essential if we are to regain our understanding of information and how it informs civic life in America.

Recognizing this new architecture as a network and embracing a little graph theory can help us make sense of our complicated tangle of relationships with the information we need in order to function in society.

We must grok, embrace, and learn to leverage this graph if we expect our media and information systems to deliver on their potential and help us reclaim our civic life.

## THE GRAPH

It is natural to understand to whom we're connected and how as a network—for example, a series of people (represented as dots or circles) in friend relationships (signified by connecting lines), as on Facebook. The idea of a social graph, even if we don't call it that, has become a pretty comfortable concept to people who use such platforms regularly. A graph is the mathematical concept for this type of network, and the construct is a useful way to reveal how our media systems really work, how information flows through them, and how we can take advantage of them.

A graph is made up of nodes and edges. Nodes are the things being connected—the dots. Nodes can represent people or companies or towns. Edges (or links) are the things that connect them—the lines. Edges can represent relationships or streets or pretty much any other kind of connection. Graphs can be used to describe all the houses in a town (nodes) and the streets used to connect them (edges) just as easily as they can describe how an alumni community of high school classmates (nodes) remain interconnected via Facebook (an edge).

Imagine the town where you live. All the people you know in the town are nodes, and the streets are edges. A graph can be used to describe all the possible paths between you and each of these people and between them and each other. That graph also could be used to calculate the most efficient way to deliver groceries. Other attributes could be added to help further describe each node—say, how many people each of your friends is connected to, or how influential each person is relative to local community goings-on. In this way, the graph could be used to determine the most effective ways to communicate information among the whole group. For example, if you know Mary makes announcements in her church, then you could tell her about the farmers market and expect that the whole church would know.

The graph is the mathematical model for representing and calculating what we all do naturally in our own minds with anecdotal knowledge. It also shows us that the individual power and influence of every node must be taken into account when we think about the flow of information, a shift in culture, or a transformative new social narrative.

Now imagine a graph that describes our media systems. The nodes are all of us as individuals plus all the institutions in our society—companies, nonprofits, government agencies—including the media companies themselves. The edges are all the ways we deliver content and communicate with one another: email, SMS, phone, one-on-one personal conversations, large-scale live events, television, radio, films, and so forth. It is important to distinguish between CNN the media company (a node) and CNN the channel we watch (an edge). Nodes that control edges are special cases in the graph because they have more power: they set the rules for how information moves, what content is allowed, and (perhaps most important) what is optimized for on the edges they control. This power is what gives a company like Facebook such an outsized influence on modern storytelling.

As the number of shared edges between two nodes increases, so does the potential capacity for the two nodes to influence each other. If I am connected to an old high school friend only via Facebook (a platform I rarely use), I may never hear his opinion or be influenced by information he shares. But if, in addition to that edge, we also connect at a conference each year, and if we share text messages the rest of the year, the likelihood increases that I will both consume information from him and consider that information valuable.

As the potential to influence increases, so do the odds that we can reliably infer something about the closeness of that relationship. The idea of proximity is essential to mapping the subcommunities in the graph, where groups of nodes are highly interconnected or share the same content repeatedly from the same sources, making their information landscape similar. The basic mechanics of the graph are straightforward. But the implications of this seemingly simple shift in architecture for how we teach, learn, track current events, understand history, and even entertain are profound.

The graph behaves differently from our old channel-based model in part because we are able to behave differently within it. In a channel-based world, our roles are generally fixed. But in the graph, all nodes share the same potential behaviors, as described earlier: each of us is a creator, publisher, sharer, and consumer of content. This increase in freedom gives each of us substantially more potential power to influence others and to be heard beyond the reach of our direct connections.

That potential power is not guaranteed, however. Relaxing restrictions on certain roles does not necessarily represent a democratization of media. There is no equality in the system. In fact, another fundamental characteristic of the graph is that the view from every node in the graph is unique. Some nodes are connected to many, many others, and this higher degree represents a greater capacity to distribute information, to be the source for more and more others. Additionally, the relationships between each node are unique, and how each node interprets information received via each edge is also unique. So while you and I might both receive a Facebook post from the same individual, I may place more value on that person as a source of information, and you might devalue Facebook as a delivery mechanism. Therefore, that piece of content is likely to have a much greater impact on me in terms of either my emotional reaction, my willingness to take action in response, or my opinion, perspective, or beliefs.

There are several components of how influential one node may be on another. The relative value we place on edges is referred to as "weight." Just as each node is unique, no two nodes have exactly the same connections within the graph or exactly the same weighting of edges. So, in mathematical terms, each individual experiences the world in a unique way, fed by institutions, media companies, and neighbors whose behaviors are also fundamentally altered by operating in the graph.

As a map of our relationships, the graph can help us understand why the world appears a certain way to us while appearing a completely different way to someone else. The view from every node in the graph is unique, but as the average degree in a group of nodes goes up and as the density of interconnectivity for that group goes up, the similarity of their experience

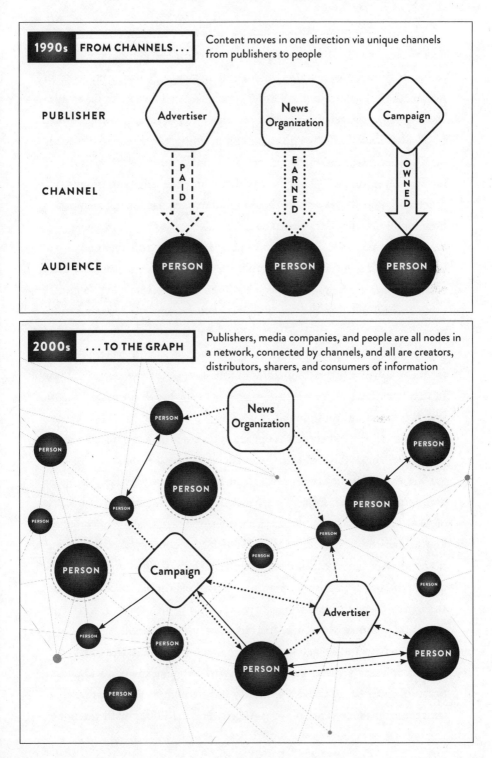

of the world converges. Depending on whether that group draws information from or cuts itself off from other parts of the graph, intentionally or unintentionally, the group may become increasingly homogenous in its information consumption, will experience fewer and fewer alternative perspectives, and the relative number of valid ideas it accepts will plummet. If the platforms that make up the graph optimize for that kind of sorting (again, whether intentionally or not), the momentum toward self-isolation increases even further. Our ability to maintain connectivity to groups that differ from us collapses until we begin to occupy the narrow, self-reinforcing filter bubbles Eli Pariser warned us about.

As has always been the case, some people maintain a broader network of connections than others. Some people are just more social. Some are more curious. Some are more intentional about exploring diversity. We all know these people, and they play an important role in our lives, opening us up to stories and information we might not otherwise hear. In the graph, they are the nodes that connect entire groups to each other and make it possible for the graph as a whole to maintain some level of *interconnectivity*. These superconnectors help keep entire segments of the graph from breaking off and disconnecting from the rest.

We often seek out certain people as authorities on a given topic. In the old model, the publishers and the channels themselves once acted as our gatekeepers on authority. The editors of the *New York Times*, for example, didn't consult the conspiracy theorist Alex Jones on politics—not because of an ideological bias, but because he is not a source of credible information. Gatekeepers of the past, however, have less and less control over distribution and discovery these days. As their gatekeeping function is redistributed in the graph (and sometimes obscured altogether), it becomes harder and harder to distinguish who is truly an authority on any topic because we lack shared markers for who is an authority or what is credible.

As flawed as the old gatekeepers were in terms of inclusion and empowerment, at least they were known and shared. Without them, we are forced to accept or use substitutes for authority and credibility. And when we combine the challenges surrounding authority and credibility with the failure of gatekeeping and interconnectivity, we start to see what Yochai Benkler, Robert Faris, and Hal Roberts identified in their 2018 book *Network Propaganda*: a system of systems where the small circles of conspiracy theory that have always been part of human society can easily explode into mass misinformation and disinformation engines.

When noncredible but emotionally resonant and self-confirming falsehoods find their way into edges where we assume authority and credibility—such as the old gatekeepers like the *Washington Post* or ABC News—misinformation and the coverage of misinformation become mainstream. Both misinformation and disinformation may spawn in the dark corners of social media, but they spread exponentially through the graph when amplified and redistributed via traditional edges like television. When misinformation becomes a common feature of our information ecology, it encourages us to retreat into narrower, trusted spaces.

As the graph around us gets denser and we repetitively consume the same story from multiple nearby nodes, stories begin to *feel* true regardless of their credibility or the authority of their authors. Proximity and frequency end up taking the place of authority; this substitution leads us to an information landscape based on confirmation bias and popularity rather than accuracy or value. The nodes most closely connected to us have profound control over what we see based on choices made for us by algorithms that treat nearby nodes as sources of content and influence, not just sources of relationship. Thus, a graph that seems designed to encourage a massive expansion of perspective ends up limiting our worldview.

The ease with which we are able to build relationships and maintain them at a distance is a key factor of this architecture. Sociologists and anthropologists refer to the maximum number of close relationships humans are able to maintain as "Dunbar's number" after the British scientist Robin Dunbar, who first suggested the concept in the 1990s—before the internet and social media began altering our social architecture. Dunbar's number is largely based on the relationship between brain size and social group size. Dunbar estimated this number to be approximately 148 relationships, and subsequent studies have generally reinforced both the idea and the number (often rounding it to 150 to keep things simple). However, the figure is dependent on a huge number of factors and assumptions about not only the neurological limits on how we create and maintain emotional connections but also the social structures and technical mechanisms that allow us to foster these connections in the real world.

While brain size (and therefore human cognitive capacity) is still evolving, that evolution happens at a multi-millennial pace. Humans have been basically as smart as we are now for at least the past sixty thousand years or so. Thanks to rapid development of new technologies, the decreased friction of our connections is changing the social structures available for us to use at a pace almost none of us can keep up with. These new tools offer us dramatically easier mechanisms for creating and maintaining weak ties. Do more weak ties suggest that Dunbar's number is increasing? Or should that concept refer only to strong ties and thus remain a relative constant? These questions are beyond the scope of this exploration (after all, I'm not an anthropologist). But certainly, when combined with ever-expanding node-to-node sharing mechanisms, our ability to maintain a greater number of weak ties fundamentally changes the number of stories and the variety of content we can readily access in a given moment. Stories sourced from personal connection (whether weak or strong) are interpreted differently from mass-published content, and so our experience of the information landscape is unquestionably altered.

One of the central promises of modern media is that more connection *should* lead to more public good. But in that explosion of connections and alongside the breakdown of institutions as gatekeepers of information, we've become subject to the constant transformation of our reality by those around us—and of truth itself. Thus, our ability to distinguish authority of information collapses in favor of proximity, which may or may not have anything to do with authority. This *potential* access to more information and wider perspectives versus the *actual* narrowing of consumption and integration of greater diversity of information echoes the paradox between the promise of modern media (*potential* social good) and the disappointment (*actual* polarization and disconnection).

As new platforms have emerged and introduced new behaviors into the graph, some of those nodes have drastically altered the basic landscape of power in storytelling and therefore culture. However, unlike in Spider-Man's case, this dramatic expansion of *power* has not necessarily resulted in an expansion of *responsibility* for these nodes. In fact, Section 230 of Title V of the Communications Decency Act of 1996 explicitly absolved some of these nodes (the online connectors and new platforms such as Facebook) from any legal liability for the content transmitted using their edges, much less any moral obligation. This failure to place the responsibility *anywhere* in society, coupled with the general unwillingness to face the consequences of that policy honestly and directly, has led to some profound negative consequences—especially for political speech and journalism. It is telling that the congressman who introduced Section 230 (Ron Wyden, now the senior US senator from Oregon) has become one of the harshest and most vocal critics of the platforms that commonly hide behind this immunity.

In a world where influence is more attainable and where gatekeeping fails, new voices rise. The rise of more and more diverse voices from historically marginalized communities has been a great consequence of this freedom. But this opening of media has also made way for the emergence and reemergence of detrimental voices that society had previously excluded from mainstream political dialogue. The shifting of and separation into various independent Overton windows rather than a shared one creates opportunity for radicalized voices to find beachheads of perceived credibility scattered across the graph and then to wrap themselves in the auspices of valid political speech.

Further incentivized to extremism by the business models of the platforms we rely on for communication, these radical voices have seized an increasingly significant role in our valid political debate. Sometimes they pretend to be credible (conspiracy theory engine QAnon for example), and sometimes they make no pretense but are explicitly unabashed in their radicalized disaffection with society (such as white nationalists). The inability to distinguish between the two adds another hugely problematic dimension to the challenge of interpreting information.

At a personal level, we tend to manage all this complexity seamlessly. I can use the graph to understand my family and how we are connected and stay in touch. The graph helps me understand that while I am connected to my mom via email, Facebook, SMS, phone, and personal gatherings, she doesn't use Twitter, so if I want her to understand what I'm talking about, I can't reference things I see there. Among all those connections, if I'm not going to visit more often, she (like most moms everywhere) strongly prefers me to call. In the language of graph theory, this means that my mom and I are directly connected by several different edges with different weights: a visit carries more weight than a phone call and still more than a text message.

Mom and I are both connected to many of the same nodes (my wife, my cousin Scott, the Obama campaign, MSNBC, and so forth), but we each use different edges to connect to them: I use Twitter and SMS, while she watches cable news. We both get content from some of the same places in different ways, and we both pass some of that content on to each other. Content she sends me from MSNBC, I generally ignore. Content I share with my mom about political conversations, on the other hand, is extremely influential over her perspective, because I work in professional politics and Mom trusts my opinions (sometimes). Different nodes using different edges to distribute, discover, and consume content—some of which is different and some the same—creates different reactions in the respective nodes. This dynamic reflects another central concept related to weights: influence.

Traditionally, in our channel-based world, we interpreted direct communication as the most powerful. For the most part, however, we were confusing *power* with what is most valuable—that is, monetizable. In many cases, *indirect* communication is more powerful than *direct* communication in terms of influence and shaping our world. A piece of content I ignore in its first transmission to me, via the *New York Times*, might entirely transform how I understand the world when shared with me by a close friend whom I consider an authority on that topic. Sharing content indirectly has always been possible (print publications have always distinguished between circulation and readership), but it now makes up a much larger proportion of our information consumption. Yet from a business model perspective, this indirect distribution is generally much more difficult for publishers to monetize—MSNBC often doesn't know when my mom shares their content with me. Distribution mechanisms that enable sharing (such as Facebook) have disintermediated that revenue, endangering traditional publishers that have failed to discover new models in this new digital era.

The greater opportunities here come with greater responsibility for us as individual participants, too. If anyone who creates content can claim authority, each creator needs to demonstrate the humility necessary to allow people to make healthy decisions about what they are consuming. Abusing the potential power of any node to wield authority and influence in order to manipulate others is offensive and problematic. Just as deeply problematic are the incentives for media company nodes to support that kind of extreme behavior because it boosts the weight and usage of their edges, and this is exactly why the distinction is important between, say, CNN the *node* and CNN the *edge*. If the economic survival of CNN the node depends on changing the rules and content of CNN the edge, then the role and value of the edge called "television news" has changed based on economic need, not information need. And that change is exactly why the public conversation about the public good needs to be reclaimed in the space of public service: to create guardrails that serve *the public* rather than serve any particular node, whether it be Facebook or a blogger or MSNBC.

This new model for thinking about the world isn't all bad news. For one thing, not only is it a more accurate way to represent our reality than the model we've disrupted and moved beyond, but it also creates new opportunities to quantify much of the conventional wisdom we've relied on for years in communications. Claims that certain things "just can't be measured" weren't true in the past and are very untrue now that the graph offers new mathematical solutions.

All these concepts have shifted for the institutional nodes in the system (brands, companies, political campaigns, governments, etc.) just as much as for us as individuals. After all, there is only one graph. As communicators and marketers, institutions must adjust how they leverage these concepts. The same power shift and dramatic increase in opportunity presented to individuals exists for companies and campaigns—as does the increased responsibility, which they also have not embraced.

If people must rethink conventional wisdom, then so must institutions. Communities, not static audiences, are the norm in the graph, so the *diffusion* of content matters more than first-order reach. The weight of an edge (and therefore its trustedness) is in some ways inversely proportional to its price, with free edges deemed more authentic than paid ones like ads, so we must rebalance content creation and delivery accordingly. This does not mean that TV is dead or that advertising doesn't work. But these are simply inputs—doses in an increasingly complex formula of influence—and are considerably diluted relative to the past. We love categorical statements: "TV is for old people . . . Newspapers are dead." They make for great blog headlines, but they rarely reflect a precise understanding of reality. Perhaps old-school media ought to be dead, but all these edges still have value and weight among some nodes for whom they are ever-changing and unique.

The value of media, in both commercial and cultural terms, has long been based on reach—how many people consume a piece of content—and on the assumption of a strong link between reach and influence. Advertising dollars are attention- and impression-centric; to maximize profit, you need to attract either a lot of eyeballs or very special ones. To drive large-scale cultural conversation and acceptance, you need large audiences to repetitively consume a key narrative.

Consider the power of a particular channel: How many people can marketers reach via television? How valuable is a Super Bowl ad? The answers rely on the direct reach of particular channels and content: Who is viewing the ad? Via what channel? But if content is easily portable—if it can be consumed in unexpected ways via unintentional edges—then accurately measuring consumption is a function of all the direct viewership plus all the indirect viewership, over some time interval. The impressions of the MSNBC content delivered to my mom directly may be influential, whereas the indirect impressions from her to me may not be. Yet if that path were reversed, the same content from MSNBC delivered by me might be *more* influential than if consumed directly. So in the graph, direct connections between nodes are equivalent to our traditional definitions of reach. If what we *really* care about, however, is total volume of influence—the weighted diffusion of content through the graph—then we must calculate not only this direct, first-order consumption but also the subsequent second-, third-, and fourth-order consumption in order to reveal how far something has diffused through the graph and how much value accrues from each step of diffusion.

From a big-picture perspective, the evolution from narrow channels to a single, massive graph has introduced a general new foundation for understanding and organizing our social world. The graph has also introduced specific consequences in the lives of every individual living in America today. By identifying these consequences, we can pinpoint exactly how we got here, why we feel so lost—and how we can find ourselves again.

## PREDICTABILITY VS. OPPORTUNITY

One key feature of our old, hierarchical, channel-based model (and one reason people hold on so desperately to it) is predictability. The stable, singular relationships and the stable, predictable content controlled by a relatively constrained set of publishers lent an air of consistency to storytelling that was especially important for commercial media. In the graph, in contrast, everyone is at once creator, publisher, and distributor. Content often flows across edges without permission and without regard to the creator's intent. Whereas people used to have only the power to change the channel, now we can redistribute content, reshape conversations, and start our own conversations at will. We can organize other individuals with relative ease, build community at scale, and rebalance power dynamics.

Brands and companies are long accustomed to wielding most of the power in our communications landscape by controlling content and conversations because they are the biggest buyers. Yet more and more often, as organic and paid content rebalance, they are being pulled (sometimes kicking and screaming) into conversations they want to avoid and forced to act as participants. This is not to say that we are all equal—far from it. Each node in the system has its own level of influence relative to each other node, depending on proximity, relative degree, and weights of the edges (direct and indirect) they share. But all nodes are (like it or not) participants in a system that is harder to control and that requires us to adopt new skills to effectively engage and communicate.

For brands, this means their behavior has to expand from broadcaster to participant—a shift in the understanding of their own identity that is often uncomfortable for or inaccessible to institutions accustomed to control. To accomplish this shift, organizations must acknowledge that the relationship has changed: they are no longer a broadcaster publishing a one-way message to an audience, but part of a community creating, publishing, and sharing content back and forth with other communities. In this dynamic, all marketing is community organizing—an exercise in building community and empowering people to share a common story, to communicate on our behalf, and to leverage our power as their own. Listening, participating in conversation, allowing others to control some of the message are all new and fundamental skills to most companies—and ones that many struggle with.

As individuals, we navigate these natural human behaviors much more easily, although they are often in conflict with our institutional instincts. Political campaigns, too, struggle with the lack of control implicit in this new landscape, despite its offer of enormous opportunity to tell stories and invite people to participate. The graph may offer greater opportunity for campaigns to organize, but to be successful, it demands a more humble, more collaborative, less controlling posture—not a natural instinct for most candidates either.

The media companies and network platforms are nodes, too, that just happen to control one or more of the edges that connect them to people or people to one another. This makes them unusual participants in the graph, as both node and edge. Their control of some content portability amid new communication norms forces on them a responsibility that they are too often loath to embrace. Thanks to their explosive growth and ubiquity, social media platforms, in particular, have become de facto public spaces, and we have handed them the role of defining how we want new communication norms to evolve and new systems to impact our digital public spaces. Determining how we want these systems to work in service of the public good should have been a public responsibility, a broad conversation between citizen and government about what we want (and don't want) and what we need from our public spaces. But out of a combination of uncertainty among the public, lack of vision among leaders, and outright desire to privatize the conversation among companies, the public ultimately abdicated that duty to the platform companies themselves and allowed rules and norms to be set by nodes whose incentives aren't necessarily aligned with the health of the graph as a whole.

We handed the henhouse to the foxes.

Media companies are (like all companies) inherently biased: their success and survival depend on how users and customers interpret and use their edges. Unless they are formed as benefit corporations (which still make up a tiny proportion of companies and none of the large media and technology companies that control much of the graph), the nature of US corporate law generally forces them to maximize profits. They are not naturally (or legally) going to hew to the public good unless it happens to align with their needs as a company. Maximizing the value, usage, and weight of a particular edge (say, Facebook) incentivizes certain user experiences and algorithms that enforce extremism, conflict, and confirmation bias over healthy perspective, visibility, nuance, and empathy. It is exactly this conversation we need to be having: How can we leverage the increase in connectivity—and the opportunity that creates for us to maintain more relationships at a distance—to sustain more—and more healthy—relationships on the back of these new technologies? How can we ensure that our new media systems function in service of *us*, not of Facebook's profitability?

Beyond the nodes and their shifting roles, a series of persistent misunderstandings steeped in conventional wisdom about our communication model make embracing our new reality even harder.

First, the gross rating point—the sacrosanct metric of media impact that underpins the traditional ad-buying cabal—is a measure of potential impact based on potential impressions multiplied by frequency and as such is not an actual measure of anything—except maybe buy size. Attribution (what content gets credit for a given outcome, like purchase or opinion shift) is simple in a closed, linear system, but incredibly hard in a massive, multivariate graph. We continue to use inaccurate first-click, last-click attribution models while talking ourselves blue in the face about the potential for mixed-media models because we can't do the math. We continue to accept the lies of reach and attribution because no one has figured out an effective way to represent the truth. This means we are reinforcing bad habits about media buying and how and where we reach people.

The second misunderstanding is rooted in the myth of unbiased journalism. News has never been unbiased; looking back far enough, American newspapers were originally owned by political parties. That bias has simply been pushed under the surface by the false cultural construct of an "unbiased news media" that embrace the same business models as entertainment channels while claiming completely different behavior.

Journalism has changed from the hyperpartisan days of the nineteenth century, starting with the establishment of new norms by Columbia Journalism School in the early twentieth century. Even so, attention-seeking business models today optimize for truth only coincidentally. Revealing their bias exposes the cultural lie that our beliefs reflect truth while our opponents' perspectives are mere propaganda; it is not a new deception, but allowing it to broadcast under the guise of a news banner is deeply problematic. Analysis is a meaningful and essential part of journalism, but modern media outlets cynically conflate opinion and commentary with news and information in order to maintain "watchability" and drive viewership, especially under the pressure of a twenty-four-hour television news format. Reporting versus commentary is a significant distinction that has been intentionally blurred for the purpose of profit, without any regard for the health of public discourse—or even worse, to undermine public discourse intentionally for political ends.

With this breakdown of stability and predictability, some of the assumptions society relies on to function effectively no longer work the way we need them to work. We need to understand what information is credible and who is an authority on what. And those understandings need to be accurate, precise, and shared, not unique to each of us. Perhaps we need different business models for different types of content whose roles in our society, culture, and civic life are different. To maximize their contributions to a healthy, vibrant society, how Disney monetizes our modern mythology in the form of the Marvelverse and the Star Wars saga may *need* to be different from how the *Washington Post* monetizes long-form investigative journalism into corruption in government.

Lacking a functional public sphere undermines the entire concept of public good, public service, and public debate in favor of narrower and narrower constituencies. But if we're willing to reexamine our assumptions in light of a new understanding, we can design a new public sphere and build a more productive culture of politics and public service. A democratic system of self-government is inherently a system of faith. For our system to work, we need to be able to assume a certain amount of predictability, so all of us can benefit from the opportunities that greater connectivity creates for the exercise of both individual and collective power.

## AUTHORITY VS. POPULARITY

In our old channel-based system, authority was generally a feature of the publishers. It was an institutional feature, something bestowed upon authors as a function of the institutional perch from which they told their stories—something we assumed about Walter Cronkite because of CBS and about all the writers for the *New York Times*. We trusted the content they produced because we understood that it met some threshold of editorial quality and validity before making it to air or print.

Some authors built credibility of their own—a trend we have seen with the rise of platforms geared toward individual participation and authority over institutional authority. Andrew Sullivan, for example, broke from

*The Atlantic* and then *The Daily Beast* to create his own media property in 2013 around his own writing, with some success. Too often, however, this individualization is more about celebrity than expertise or authority. In cases of less credible content than Sullivan's work, this often makes for dangerous confusion.

The idea that what an authoritative gatekeeper selects for us is inherently valid is a major assumption that goes back to indigenous oral cultures and a time when we implicitly believed the stories told through our elders—and one that has not survived the rise of the graph. When what we consume is entirely (or even partially) based on who is willing to pay to present us with content, or on what the people near us in the graph consume, that assumption of authority fails. Instead of an authoritative source, we are presented with information in service of commercial interests—whoever can afford to buy our attention or whatever is popular. Some of that content may contribute to healthy democratic debate—to being well-informed or prepared to participate in an active civic life. But next to none of that content is optimized by these media systems for healthy debate and credible information.

Whether or not those gates were ever valid, they no longer fully protect us from content coming from anywhere and everywhere. The more we interact with content through streams that mix information from different sources and make everything appear to be of equal value, the more we are completely on our own as we try to determine authority and judge which stories are credible.

There is a growing awareness that we are increasingly required to question the authority and validity of the things we see and read. But we have been taught terrible habits by platforms that do little to help us distinguish credibility and authority, and we lack the skills to assess content effectively. This individual-level failure is an important element of this credibility collapse. Regardless of the bad intentions and misaligned incentives of the platforms and corporate media providers around us, we are failing ourselves, too—by not effectively improving our media literacy, building our capacity to decode sources of authority and credibility of content, questioning our own assumptions, or confronting our confirmation bias.

## CREDIBILITY VS. CONFIRMATION

Related to questions of authority and, ultimately, trust is whether the content we consume is credible. Credibility is a content feature affected by a matrix of conditions, including who we are (recipient node), which node the content originates from (source node), our relationship to that node (node weight), which edge delivers the content and how (edge frequency), and the weight we give that edge (edge weight). The solution to the matrix is unique for every person, every piece of content, and every goal of that piece of content.

In the early days of social media, people were obsessed with the concept of influence and credibility, and many attempted to calculate influence as a generic attribute. But in reality, influence is unique and context specific. Dr. Anthony Fauci and President Trump, for example, have different credibility (even among Trump supporters) when it comes to the coronavirus pandemic compared with something less personally risky like trade policy. Our emotional frame and our sense of risk inform our assumptions on credibility, and when credibility is unclear, we tend to substitute other emotions or concepts (such as proximity) out of desperation. The murkier credibility becomes, the more desperate our assumptions and the less reliable our relationship to the world.

If we are to rely on the credibility of our storytelling infrastructure, and for it to drive healthy civic conversation and participation, our modern media must help us build a trusted view of the world around us—and that view must be shared, at least in part, by others. If the field of trust and shared trust narrows in the same way our field of information has, our ability to communicate effectively beyond our like-minded segment of the graph fails. And without a functional model for credibility, we substitute confirmation bias for trust, and our ability to span communities fails as well.

The elimination of the traditional gatekeeper architecture of twentieth-century media has freed up new voices and unmasked a greater diversity of perspectives. To fully embrace this new reality, we must discover new shared mechanisms for assigning a shared understanding of both authority and credibility so that trusted conversation can become a consistent feature of modern media. Trusted conversations require authoritative voices and credible content, not unreliable influencers and self-affirming feedback loops. If democracy is indeed a system of faith, then trust is an essential element required for it to function.

## DIRECTED VS. UNDIRECTED

In our comfortable, channel-based view of the world, content flowed in one direction (from publishers to audiences, via media companies or platforms) and generally only in first-order ways (directly from media company to audience). Most content was not portable: it could be discussed but not easily shared. In oral traditions, people had to hear stories over and over and over before they could reliably share them. But with the undirected edges on today's graph, content readily flows in both directions. I can text my mom, and (whether I like it or not) she can text me back.

Twitter feeds. Facebook walls. Comments sections on blogs and websites. More and more of the mechanisms by which we consume information are undirected and invite response. We have traded passive consumption for active conversation as the standard posture of storytelling. This shift is so much the norm that we are often surprised when we don't have an opportunity to respond. As conversation mechanisms rather than broadcast mechanisms, they demand the level of consistent engagement and participation that a conversation requires.

Organizations and institutions that rely on mass communication—from brands to political campaigns—are used to being the storyteller but not having to participate in the conversation, so they often learn this lesson the hard way.

As a building block of our community in the 2008 campaign, Obama for America stood up our own social network called My.BarackObama. Anyone could create a profile, connect with other supporters, start their own blog, and lead conversations about whatever was most important to them. It was a powerful statement about who our campaign was for and how we were willing to devolve power down to our grassroots supporters. We didn't realize immediately that the hundreds of thousands of people and conversations needed us to participate as well, that our role included validating the value of the experience through our participation and hearing what mattered most to people. Senior staff ended up having to wade into complex political arguments—like Denis McDonough, who would become deputy national security advisor and eventually White House chief of staff, holding a public comment session on closing Guantanamo Bay. It took us time to recognize that our community demanded we participate in the experience we created for them.

In 2010, United Airlines broke a passenger's guitar that was traveling as checked baggage. It was nothing intentional or malevolent, just a simple mistake. Looking for some sympathy, and possibly hoping for some customer service, the guitar's owner took to Twitter and then Facebook to connect with United. The company didn't reply, even though it had joined these online spaces ostensibly for exactly this kind of communication. When United didn't hold up its end of the conversation, the guitar owner turned elsewhere, looking for a connection. The owner took to YouTube with a catchy song about United's failure and became an internet sensation, embarrassing United on a massive scale. By failing to recognize that these platforms were not just more broadcast channels it could control, United drove its customer to louder and louder behaviors in order to get a response. The company's simple mistake snowballed into a massive misunderstanding—a harsh lesson about the difference between a broadcast channel and an undirected edge of the graph.

## POWER VS. REACH

Storytelling is about more than the transfer of information or history; it is also about the setting of norms and culture. What we know about the world, what we believe, and what inspires us to act—these things shape how we behave. So storytelling is also an expression of power.

In our old model, reach was naively equated with power. Now, the potential power of any piece of content to inspire, entertain, persuade, or activate is a function of *how far* it moves through the graph combined with *how* it moves through the graph. Volume represents the total diffusion of content through the graph. It is a combination of overall first-order reach, plus paid impressions (if any), plus impressions created through sharing to nodes of the second order (and the third, fourth, etc.). In this way, diffusion is the graph's version of reach. Mathematically, power can be calculated as follows:

$$\textit{diffusion} \times \textit{average node weight} \times \textit{average edge weight}$$

Power produces a status change in a node or set of nodes—a conversion from some state A to some new state B. Giving credit for that status change to a particular piece of content is the process of *attribution*.

If I've never heard of, say, a new movie that is coming out soon, and I hear an ad on the radio for it and become aware of its impending release, I have converted from a state of unawareness to a state of awareness. If that ad is so compelling that I intend to see the movie, I have gone from being unaware to being an intended buyer as the result of a single piece of content. Now, if I had seen that ad on Facebook and it had offered me a link to click and immediately buy a ticket, I might have gone from unaware to purchaser in a single conversion.

Generally, we go through these steps one at a time: I learn about something from an ad. I hear more about how great it is from a friend via a tweet. I hear another ad on the radio a week later. Then I read an article in a magazine about one of the stars of the movie. And another week later, when my wife wants to go to the movies, I say, "I want to see Christopher Nolan's new movie *Tenet*—it sounds amazing." ("You want to see all of his movies," she would respond. And she would be correct.)

At the end of this chain of communication is a commercial transaction that the movie distributor has set up to get me and my wife to buy tickets to this movie so the company can afford to keep making and distributing movies. And it worked. Each piece of content—created by different sources, delivered to me through different mechanisms—had some amount of power over my thinking and behavior relative to that movie. The attribution question is: Which piece of content gets credit for me buying the ticket?

The advertising world typically assigns credit to either the first ad or the last ad in the chain that leads to the buyer's ultimate conversion. The problem is, it's a dramatic oversimplification of the incredibly complex set of micro-influences all around us, all the time. A more accurate accounting would include the small expressions of power in hints and nudges that all add up to an opinion, a belief, and finally, a choice.

When we start to think about storytelling and communications and the expression of power in civic life, this complexity introduces huge opportunities to more effectively engage and organize the world around us. Since all of us now have the capacity to create and distribute stories, we are all more empowered to express our individual minds. And the potential for leadership gets decoupled, at least partially, from institutional position. But in modern political campaigns, we leverage this chain of advertising at an unprecedented scale, spending billions each cycle on paid media in an attempt to drive voter opinion and behavior. A huge proportion of the debate about campaign strategy has centered around the mix of tactics: How do we leverage television, digital, and mail? How do we balance media and field? Yet an honest analysis of the power of all that paid media that takes into context a more complete equation—one that considers influences and the tremendously varied environment each voter occupies—might call into question the effectiveness and efficiency of our strategies and suggest a more integrated view of how each edge relates to leadership.

If we look at early polling from the 2016 election cycle and the final outcome, there was very little movement in support for Donald Trump versus Secretary of State Hillary Clinton. Although we most often attribute this to the hyperpolarization and stratification of political perspectives in American politics, it is also possible that the billions of micro-influences that shape voter opinion are dominated by cultural signals we mostly

ignore, but that ultimately outweigh all the efforts of the campaigns themselves. Campaign strategists wave away what is driving people's perspectives, offering excuses for what has made our strategies ineffective; they call
voting blocs immovable, when in fact certain voters simply are not subject
to movement by campaigns directly or by traditionally favored tools. It
may be time to rethink the chicken-egg question about culture and politics. We have to reimagine what it means to campaign in the graph.

## COMMUNITY VS. AUDIENCE

In a channel-based world, the core question for content creators or marketers is: Which channel will provide the largest audience? And the audience's
role is to consume what is passively presented. The model is simple, linear,
hierarchical. In the graph, via undirected edges, we talk back. We respond.
We redistribute. We comment. All content ends up being a conversation,
whether the creators like it or not.

The platforms that have done the connecting to get us where we are
today, knitting the fragments of the 1980s and 1990s into this messy graph,
were primarily designed as relationship platforms. Their origins are a clue
to their best use; they were intended to help people build and maintain
relationships at a distance. The idea was to bind mass groups by common
interest or purpose in relationship: community. The outcome, however, is
a collection of similar people consuming similar things, yet disconnected
from each other.

This shift in group dynamic also means that as storytellers, we must
be prepared both for asynchronous consumption and for immediate commentary that produces even more content than our original story. And as
institutions, we must consider all communications an aspect of community
organizing. People's expectations are no longer passive.

In the graph, power is more widely available. All content has the potential to create action, to distribute power, to inspire new behavior. The more
we are able to reconceptualize communications as an expression of power
and an active invitation rather than a passive, one-way delivery of message,

the more effective our storytelling becomes. And storytelling that is meant to be part of a relationship also encourages better balance of power and more agency among individuals in politics. Rather than being shouted at and told what to believe, individuals might be engaged as powerful members of a community invited to a dialogue. We are encouraged to take a story back into our community, to make it our own. We are given the choice to become part of the process of leading the community.

The shift toward community, combined with the consequences of tapping into directed and undirected edges, creates the conditions that ought to enable more and more vibrant civic discourse—an always-on opportunity to participate in a dialogue about who we are and what we need from our leaders and for our communities. Our ability to debate in safety, to live in perpetual dialogue, is fundamental to a healthy, productive civic life. Self-government is not a system of solved problems but one of constantly solving problems, and that constant process requires constant dialogue.

The unpredictability of the graph and the collapse of authority and credibility make establishing the trust necessary for this dialogue complicated at best and impossible at worst. The potential for exactly this kind of engagement is, in fact, inherent to the architecture of the graph—but the misaligned incentives we have embraced and the commercialization of that architecture have hijacked that potential.

## CHOICE

Individuals have never had much power when it comes to the stories that define culture. Whether handed down through elders or imposed by a priest or king, our reality has generally been fed to us in the form of the dominant stories of the day. As channel-based media fragmented at the end of the twentieth century, however, we started to develop more power of choice. We could simply tune out certain stories, and even those in power had to work harder to acquire and retain our attention. But because the view from each node is unique and determined by the behavior of the nodes around us, we now have even less control than ever over the choices in our view.

Culture is crafted, and history is written, by those who tell our stories. As the creators, distributors, and consumers of our stories change, our culture changes. Introducing largely corporate gatekeepers into the core architecture of storytelling also introduced a powerful mechanism for reinforcing the flawed cultural power structures of the past few centuries. The channel-based model institutionalized a predominantly white male hegemony over information and American culture. With this reinforcement of colonial power came the responsibility to act as a healthy gatekeeper for fact and fiction, to hold other power centers (particularly corporate ones) accountable, and to ensure appropriate distinctions between the types of stories society needs to function—information, entertainment, history, myth. Yet these corporate gatekeepers have never been willing to hold each other to account and have not acted in the interests of our collective public good.

In the graph, this power dynamic fully breaks down in favor of individuals who get to participate, at least theoretically, as equals to other nodes. In practice, each node's gravity and valence is its own. I am still not equal in power to CNN, but CNN does have to work a lot harder to get me to use its edge than it used to. Likewise, the White House has to work a lot harder these days to draw attention.

The problem with attention defining so much of how power functions in the graph is that attention—rather than truth—is the primary driver of value. When commercial conversion is the dominant goal and when attribution of change is driven by how we monetize attention in the form of advertising, we are trapped in a set of optimization algorithms that encourage certain behaviors and gradually diminish proactive choice. And just as the commercial gatekeepers failed in their responsibility to the public, so too are the platforms populating our graph.

They are failing us and leaving us with an impossible task.

When we're talking about entertainment, the consequences might not be as profound, although I would argue that the narrowing of cultural entertainment is not only sad but also weakens our cultural foundation. When we're talking about conversations for the public good, however—content from community leadership and journalism—the consequences are profound. A fourth estate whose business model is optimized for attention will only coincidentally optimize for the truth, no matter how well-intentioned and talented the reporters and editorial staff. Leaders focused on the attention their outrage can buy or grab from extreme communities will shift their language and eventually their values toward those communities. And the rise of algorithmic content display erodes the power of individuals by making choices for them behind the scenes without their permission.

In some cases, our initial stream of content is largely defined by our first choices of followers, like the blank slate of a brand-new Twitter account. If we excel at and are committed to actively curating the lists and streams we consume, the tools of modern media can help us build a broader perspective and extend our thinking. But we are working against the momentum of these systems. As they strive to personalize our experiences, they compel us toward a more homogenous view of the world. This disguises control as the appearance of choice, and that's where hyperpersonalization shifts from feature to bug.

When only two major companies, Google and Facebook, make most of the decisions about what information we see—and we don't even understand that others are determining our frame of reference—the power of choice shifts. It moves out of our hands, transfers to our closest nodes, and ends up with whoever controls the edges that define our content delivery (and the people who buy our behavior from them). This trend drives isolation and the breakdown of any concept of a commons. If our frame of view is unique, and then it is further optimized for us individually, we share no common concept of the public good. When the health of our body politic is absent from the algorithm, we are driven further from each other, no matter our hopes for the opposite. For those of us looking to maintain connections more broadly, we are swimming against a multibillion-dollar stream.

In oral traditions, everyone in the tribe shared the same stories. As media systems evolved from the printing press to the radio to the internet—and as society morphed from a single, stable audience into a complex, infinitely changing graph—we slowly began to share fewer and fewer stories and therefore less and less culture. Now, in the graph, we don't have to share anything at all. We can live fully realized, culturally rich lives without intersecting with the media and information worlds of our closest physical neighbors. Without either a single node or trusted edge that we all use, we utterly lack a safe, reliable public sphere, and even the potential for healthy dialogue slips away. Without this healthy dialogue, and without some concept of shared trust and authority, our democracy devolves to the mean, partisan reality we now experience. A healthy civic life becomes harder and harder to reclaim.

The actual math of the graph is beyond the scope of this book, but the concept of the graph provides us an essential frame for understanding our cultural and civic dysfunction. The graph puts a *why* to it, helping us crystalize how that dysfunction came about and what continues to fuel it. The base architecture itself has already had profound consequences for our consumption of information and the creation of our culture. As the world has struggled to catch up, companies have developed new ways to monetize the new systems. And the business models that drive both the graph and the heavily monopolistic economics of twenty-first-century America are adding gasoline to the inferno.

# 3

## IT'S NOT THE CRIME—IT'S THE COVERUP

THE PREVAILING NARRATIVE about media and information in the age of social platforms portrays an evil attention economy driving the hyperpolarization of America, ad-based business models destroying public discourse, and journalism losing an unfair fight with technology companies. While some of that narrative feels consistent with our experience, it oversimplifies key realities about the economics of modern media. Clarifying the economic reality will help us understand what's not working and point us in the direction of public progress.

The economic incentives that have provided the fuel for the growth of modern media also spur design choices and metrics for algorithmic optimization that ultimately shape the media's impact on society and on us as individuals. The failure of those incentives to push society in the direction of a healthy public sphere—in the direction of progress—leaves us abandoned and isolated as we try to find the way for ourselves.

Basic economic models are based on the principle of "perfect information"—that for markets to function in healthy ways, all actors in the system must have perfect and instantaneous information. Likewise, a free, robust press is essential for democratic self-government to function properly. The principles of a free press as a necessary pillar of functional democracy are righteous and real. We need healthy information sources that allow us to make informed choices about leadership and life. And the people in whom power is entrusted need to know and understand what is going on in society and how their choices affect us.

In economics, the principle of perfect information means that all market participants have the *same* information all the time. In a democracy, when citizens all have the same access to the same information all the time and enjoy the same capacity to understand and exercise power, that might reflect "perfect political information." But the journalists we've trusted to create and disseminate much of that information generally do not have perfect information as a goal. True to their partisan roots in the eighteenth century, newspapers are now "attention partisan": to stay in business, they must optimize for attention, not truth. The incentives that drive their business and the needs of perfect democratic information may align accidentally but are often at odds.

The economics of media have different consequences for different types of content. Information, entertainment, history—each has a different purpose and function in society, and each is motivated by different incentives.

Information is the basic building block of everyday content and informs us of what's going on currently in society. This bucket certainly includes formal news, but it also contains less formal storytelling such as gossip, which has always shaped our worldview, and opinion-based content such as editorials and analysis.

Entertainment is our container for content intended for pleasure, even if it sometimes blends into analysis or interpretation (satire) or into culture setting (history). Before literacy was widespread, theater was an indispensable method for delivering histories. Live-acted content took many forms, just as daily printed and delivered content might be news and might be entertainment.

History is the sum of the stories, generally determined by the victors, about where we've been and how we got to the present. Traditionally, history has been free to access but controlled by the prevailing power of the day: first by elders in oral indigenous traditions, then by priests, and then by kings. That control has been a key to maintaining the command of national narratives, which set the basic frames for our understanding of norms and expectations and of power and consequences.

It is important to distinguish the type of content from the medium of delivery. The blurring of these lines over the past fifty years has been part of the unsettling of the economics of media and the unmooring of our grasp on how to process what we consume.

## HOW DOES HE KNOW WHERE WE'RE GOING?

We know because we have been going in the same direction since before the beginning of the republic. Information—news, current events, gossip— is the content most germane to the conversation about the effects of media on civic life. How we understand our society, who is in power, who represents us, what they are doing in service of the people they represent, and how our decisions will affect our communities are all wrapped up in how we comprehend what's going on—and in what we think and feel about it. These kinds of information have always been social, mostly open, and highly unreliable. The acceleration and ease of social sharing has been a hallmark of the migration to the graph, but social media did not invent this behavior. Social sharing has been a fundamental pillar of the distribution of information from its origins.

In America, it started with the informal, word-of-mouth information networks of the colonial era, when our basic access to information came from traveling merchants returning from wherever they'd been with stories of events—incomplete and without context, details, or certainty regarding the source. All information was essentially gossip. With little repetition and almost no literacy, validating information was close to impossible. Consistent engagement in current affairs was virtually impossible for the average citizen, even if it had been desirable. Because of its source and method, we consumed it with healthy skepticism.

The expansion of primary education in the nineteenth century created an explosion of addressable readers. Starting with America's battle for independence—a kinetic and cultural war both internally and with the Old

World—the pamphlet became an essential tool in the ideological propaganda campaigns that ultimately defined our nation. During the First Party System era, from about 1792 through 1824, the newspaper (distinguished by its regularity) displaced the pamphlet as the mainstay of communication with voters. Owned by political parties, these early papers were deeply partisan documents meant to persuade their readers and inculcate a specific worldview. From the beginning, all of this information was generally free; a very small subscriber base shared the papers widely. These initial experiments in journalism had to acknowledge that most people would not (or could not) pay for access to this kind of content. But thanks to large political subsidies, this era saw the first attempts to expand coverage and shape understanding of a broader spectrum of events—political, cultural, and economic. In these early days, the value of being first became clear. With so little information density and so few choices, a single story might be a person's entire window into an event. Anchoring their understanding in your worldview was a massive strategic advantage.

The 1830s brought the emergence of the penny press. Cheaper, faster printing combined with guaranteed, rapid delivery by the postal service brought content to the emerging middle class. Without the subsidies of political parties, these operations needed wider circulation to generate the attention inventory necessary to sell advertising placements. So, in an effort to ensure the largest possible audiences, they shifted away from partisan postures and political stances. Gossip, scandal, and sensationalism reigned, signaling the birth of what would later be dubbed "tabloid journalism." What we saw now was a newspaper industry that viewed the medium as inherent to its product rather than simply a distribution mechanism for its content. This assumption closely and durably tied distribution to business model in a way that has resisted imagination and innovation ever since, leaving the entire industry susceptible to disintermediation—the loss of control over distribution mechanisms to faster, more scalable technology platforms.

Even as information was professionalized and institutionalized over time, sharing physical papers remained a substantial factor in the flow of information. Peak newspaper circulation as a percentage of population in

America reached only 35 percent of US households back in the mid-1940s, and that figure has declined ever since. Today, new sharing mechanisms allow more people than ever to reference and therefore consume content from major print news outlets, but those outlets cannot effectively monetize their entire audience. The acceleration of sharing as a primary or even dominant mechanism of distribution (and therefore discovery) has created business challenges for publishers whose first-order circulation (people who subscribe directly) and first-order attention (people who get information directly from the publisher) is the only means of generating revenue. Their reach has increased, but their revenue has not.

By the turn of the twentieth century, with all forms of storytelling beginning to compete in the same sizable market of attention, the advertising-based business model was fully entrenched. This model required attracting the greatest possible number of eyeballs and selling the most possible units. The options to achieve this goal increased around the turn of the next century, through the expansion and fragmentation of media; likewise, our total consumption of media has expanded but also become more divided. At the same time, we've held on to the expectation that news should be cheap or free, which has only increased the economic pressure to maintain attention to fuel advertising revenues.

Looking beyond advertising has reignited the conversation about subsidization—from either wealthy benefactors (Jeff Bezos for the *Washington Post*), partisan funding (megadonor Robert Mercer for Breitbart), or readers via subscriptions. But information had always been partisan, whether the source is a biased merchant, a gossipy and opinionated neighbor, or a political party. With the rise of "yellow journalism" (tabloid sensationalism and unsourced, crude exaggeration) in the late 1800s, many newspapers gave up their position as the primary source of authoritative information. In order to reclaim their place in society, the institutions of formal news sought to standardize and institutionalize journalism by wrapping themselves in righteousness and principle. Horace Greeley's *New-York Tribune*, followed by Henry Jarvis Raymond's *New York Times* and then Columbia Journalism School, codified the principles of unbiased, professional journalism that still anchor our foundational ideals regarding information. But this cultural narrative of journalism has always been at odds with the reality of the news business.

The real driver of reduced partisanship in the newspaper industry was not a principled moral stand, but an economic reality. As choices have exploded and attention has become more precious and expensive, journalism has found itself competing with other forms of content. With only one albeit ever-expanding market for attention, news and truth have to compete with fun, joy, and conflict. Over time, newspapers have fallen deeper and deeper into the "if you can't beat 'em, join 'em" trap, blurring the boundaries between information and entertainment in order to compete for attention with cat videos. To ensure enough attention to survive, they increasingly rely on sensationalism and conflict at the expense of nuance and honesty. James Fallows exposed the compromises and consequences for journalism and civic life in his seminal work, *Breaking the News*, published in 1997 before any of the graph architecture was in place and a decade before social media began to reach ubiquity.

In the graph, however, even those compromises are insufficient to economic survival. Revenue has largely shifted from the publisher (e.g., the Tribune Company) to the distributor (e.g., Facebook) as technology companies have insinuated themselves between publishers and their readers. Newspapers have made this easier by not distinguishing content from delivery. Despite the standard narrative portraying twenty-first-century social media innovations as offering openness and community for all, the reality is that these platforms have instilled an even more partisan, more sensational algorithm driving the same old business model. Facebook, Twitter, and Google did not invent a business model. They simply disintermediated existing publishers and effectively consolidated the market for attention-based revenue in a new architecture that made direct consumption less profitable. This disintermediation has muddled our forms of information to the point of near indistinguishability—and left the entire news industry at a loss for a new method of subsidizing the high-quality information democracy requires.

A functional democracy is increasingly impossible in a world where cat videos and news compete for our attention based on the same economics. Although the deep, investigative report on administrative corruption at the Justice Department from the *Washington Post* just might not be as fun as the video of the cat asleep on the Roomba, we need it more.

But media economics has always been an attention economy, so why do things feel so dysfunctional now?

It's the coverup.

As the graph has grown, and as we have all become creators and publishers, the amount of content available in the world has grown exponentially. From the internet's very beginning, we needed new systems to discover and sort this proliferating universe of new stories—to order the chaos and make the graph accessible. The rise of the web was driven by directories of sites to do just that. Yahoo, an acronym for Yet Another Hierarchically Organized Oracle, began in 1994 as a site called Jerry and David's Guide to the World Wide Web. As these tools grew into businesses and as the exponential expansion of the internet drew commercial and capital interest, these young companies needed to develop a business model to support their free directories and search engines. But rather than building new revenue models to fund new information behaviors, they saw an existing multibillion-dollar advertising market for attention and realized that revenue could be theirs.

The great business innovation of these platform companies is building a brand narrative depicting openness, community, and the techno-utopian mythology of greater connectivity while simultaneously developing platforms that provide each unique user with a discrete, hyperpolarized experience. They are still selling attention. They are still providing biased experiences. But they don't have to create the content. Instead they merely reveal content from publishers to users based on algorithms tuned to the user's biases. They have replaced the publisher's bias with the reader's bias—thus taking a clear, specific, shared biased experience and replacing it with an opaque, generalized, per-user biased experience based on our unique view within the graph. So an information discovery and distribution mechanism that was understood to be biased and was interpreted with that in mind—carefully, skeptically—has been swapped for a faster, more efficient mechanism with a hidden bias provided by a brand narrative that suggests safe, community-focused experience and therefore does not raise our natural, healthy, skeptical information-processing skills.

*The innovation is the obscurity.*


## USERS VS. CUSTOMERS

When it comes to today's media platforms, we may be their users, but we are not their customers. Platforms are multibillion-dollar businesses built on selling the attention of individuals (their users) to advertisers (their customers). In a traditional media transaction, either a consumer pays to consume the content, or it is provided free to her with an advertiser paying for access to her attention. For information, especially content that can be delivered digitally, subscriptions and various forms of reader-supported content models exist. But the dominant model by far is the advertising-supported model—the same business model used by newspapers since the 1830s with the invention of the penny press.

But two important things have changed in the last decade.

First, almost all the platforms that we rely on to access and discover information (e.g., Google) and that publishers rely on for distribution (e.g., Facebook) are free to use and have managed to capture most of the advertising revenue that publishers used to rely on. If a *New York Times* article is shared on Twitter, Twitter gets the revenue from the ads that appear in a user's stream around that article. Publisher revenue-sharing is beginning to happen with some of the larger platforms but is far from the norm and a tiny fraction of the revenue. And those arrangements are certainly not sufficient to make up for the *New York Times* losing direct access to its audience. Furthermore, a newspaper is roughly a fixed container that must be filled daily but limits how much content can be delivered and therefore requires extensive curation and prioritization. One can argue about the validity of that prioritization—whether it might be prioritized for civic discourse as in the *Washington Post* or for sensationalism as in the *National Enquirer*—but the limit is important. It represents and defines an end point to consumption in comparison to an infinite digital content stream.

Second, to increase the amount of content the platform can show you, and therefore the number of ads it can sell and display, these platform companies have created stream-based interfaces that not only achieve a dramatic increase in the amount of content consumed, but also make all content appear equivalent. A stream is infinite—and that is exactly how Facebook and Twitter want your consumption to feel. With minor, easy-to-miss distinctions, all tweets look like tweets, and all Facebook posts look like Facebook posts, whether they are sponsored by a company or an influencer, whether they are technically an advertisement or not. Enabling the public to differentiate content types or sources would make the stream less continuous and create more design friction that may reduce attention inventory and, ultimately, revenue. The decline in our ability to distinguish what we're consuming is an active design choice to support the platform economics—not a mistake.

Our inability to effectively distinguish content types, which is implicit in the design of the content stream, has changed the way we interact with information. In a conversation with another person, I know who I'm talking to and can make appropriate decisions about their credibility. If someone hands me a copy of a newspaper, I know who gave it to me and what I'm reading—and because the newspaper is a known publishing entity with a track record of certain quality or bias, I know *how* to read what I'm reading. Conversely, in an online stream, the content from every source appears with equal weight (notwithstanding very small distinguishing factors that are intentionally easy to ignore and hard to customize). A story from my mom looks like a story from the *Washington Post* looks like a story from an anti-vaccination blogger. This innovation is excellent for the anti-vaxxer, as it gives him distribution outside the bounds of traditional gatekeepers and erodes our ability to determine his credibility.

Healthy, skeptical content consumption has become harder and requires more work. Our old systems did much of this work for us, but we paid a price in that our content and sources were much more restricted. With the expansion of voices and choices within the graph, however, we've lost context and the ability to distinguish types of content and sources. That leaves each of us alone to set our own standards for credibility and authority. This context collapse accelerates and anonymizes the social sharing behavior that has always been at the root of information exchange and how we learn about the world around us. More information is available, but it has become harder to consume the information in healthy, productive ways.

Our experiences and communities within these platforms are designed not based on what constitutes a strong, healthy society or a united, more vibrant community, but on what ensures the most attention inventory to sell to advertisers. Platforms are designed to exploit us (their users) at every turn, making usage addictive in order to keep our attention and thus build inventory for their customers: the advertisers.

Meanwhile, these platforms have positioned and sold themselves to users as the provider of one thing that all people want: belonging. They provide us with free, convenient access to our friends and families near and far—the ability to stay in touch with more people we know and to meet people we never would have met otherwise. Yes, the number of weak ties we can sustain is going up, and greater connectivity *could* give us greater visibility into more communities—if the platforms optimized our experience with that outcome in mind. But they don't, because their businesses must optimize for attention. And what generates the most attention? Bias and outrage—not greater connectivity to greater diversity of thought, experience, and perspective.

So yes, we get lots of content for free. We get lots of access for free. But the costs of those free products are largely obscured from view; they include the transparency of bias and intent, which undermines our ability to properly distinguish content and consume information.

Furthermore, in these platforms' deference to their true customers—the advertisers—users have lost control over our privacy and our power of choice. There are way too many of us rage-clicking away for our streams to be curated by a person, so optimization is mostly done algorithmically. There is a formula at work, and that formula determines the order and prevalence of what appears in our stream and what disappears from view, based on as much data as the platform can match to each individual user. Age. Gender. Location. Purchase histories. Internet surfing habits. To retain users' attention, platforms use our information to ensure that every item in our stream is exactly what we want to see there.

Some users may be thrilled. Some may be surprised and even a little creeped out by these "psychic" powers. How does Instagram know about that particular pair of New Balance cross-country spikes I've been craving? What does it mean that the selection was made *for* me? What is the consequence of not actively making my own choices?

## EXPLOITATION BECOMES A FEATURE, NOT A BUG

Hyperpersonalization makes our lives easier. It reduces the friction between us and the content we want and the purchases we make. But it also fools us into believing that the world is simpler than it really is. It presents *apparent* choice but not *actual* choice.

Making choices is an important way of differentiating elements of our world and setting the boundaries for how we understand its scope. Even if we make the same choices every day, we are selecting from a massive set of options that describe a massive field of valid choices. Surveying that field helps us understand all the perspectives at work in our society. In political terms, that field of valid choices defines the Overton window; it helps us locate our political positions relative to the rest of the civic polity in which we participate.

These choices matter. They keep us connected to others with whom we disagree, even if it is by exclusion. Saying "I want this, not that" at least acknowledges the valid existence of "that" in a way that helps shape our worldview. Being compelled to sort, to curate our own experience, forces us to continually recheck what we believe. When media platforms sort us according to confirmation bias into homogenous segments of the graph and then feed us content that we are likely to agree with, they are tricking us into believing that *our* ideas are the *only* ideas—that everyone agrees with us. And if my information stream suggests that the whole world agrees with me, then the whole world must want what I want. There is no reason to debate. Anyone who disagrees is deemed "other," not only wrong, but at best, nonsensical—or at worst, evil. When the alternatives to what we agree with are outrage-seeking and conflict-oriented, we are fooled into believing that anyone we disagree with is out to harm us.

Not having the opportunity to choose is what makes our filter bubbles so insidiously powerful at undermining our capacity for debate. For media platforms, however, fulfilling our confirmation bias and making us feel right is a better way to keep our attention. Confronting people with countervailing opinions, forcing them to think and choose, is cognitively more expensive and therefore worse for the platform business model. But it turns out that optimizing for attention does not optimize for healthy discourse or the kind of heterogeneous conversation required in a functioning democracy. Our media and information systems run on a model that is directly opposed to what is good for our civic life.

Our civic life is constantly undermined in favor of revenue for platform companies that need user attention as inventory. They aggregate and exploit our attention for the revenue potential of other companies—at the expense of our understanding of the world, at the expense of our connection to others—and both the bias at work and the consequences are largely hidden from users' view. We didn't even realize it was happening. Most of us still don't. That's the power of the coverup.

As the Silicon Valley machine has matured, this exploitation has also matured from something hidden to something featured for its end customers. Information platforms like Facebook and Google exploit our attention for profit while carefully hiding that exploitation behind intentional design choices and a duplicitous brand narrative. Facebook is optimized for togetherness only if that includes people who agree with one another and excludes everyone else. And providing free assets—whether attention or work—in trades where the costs are obscured has now become a core feature of some of the most dominant third-generation internet platforms.

Uber is a transportation company that does not employ drivers or purchase any vehicles as part of its infrastructure. Instead, the company exploits driver time and resources by promising freedom. The exploitation of labor is presented as a benefit, and indeed drivers are grateful for the freedom and opportunity to work whenever and wherever they want. Meanwhile, the company can avoid taking care of or making any commitment to its labor force and can push all the costs of delivery from its balance sheets onto its drivers without sufficient compensation.

This model is popping up everywhere. Airbnb is a hospitality company without property. TaskRabbit, Fiverr, and Figure Eight all offer various professional services without employing professionals. There is no question that some of these platforms provide value and convenience, but the extraction of value is inarguably skewed heavily in favor of the platforms at the cost of their users.

In the industrial economy of the past hundred years, labor exploitation was seen as a somewhat inevitable bug to be fought and regulated against. Today's gig economy is an entire system built around exploitation as a core feature. Unions and fair labor standards, pay equity, and child-labor laws all emerged to push back against industrial systems whose goal was the most efficient allocation of assets (including labor) for the creation of the maximum profit. Following in the footsteps of social media platforms, the gig economy companies have taken an old model, obscured the costs, and called it new. They have built masterful narratives around empowerment and freedom while extracting maximum value from participants and sharing the minimum possible.

While commercial exploitation of our human instincts is nothing new, the relationship between neurobiology and commerce has generally been confined to entertainment (casinos, I'm looking at you). The imbalance of computing power and big data versus human instinct and psychology is profound, and yet technology is being used against us unapologetically— not just to personalize experiences, but to dictate them and control our behavior. As Levitin's *The Organized Mind* reveals, using our psychology to design interfaces that are intentionally addictive is central to the business models of social media platforms. These platforms optimize interfaces not for our benefit, but to maximize our consumption and engagement in order to expand the attention inventory they can sell to advertisers.

Compelling our attention isn't really new either, but combined with shorter feedback loops and single-click purchasing, this attention management strays into behavioral control. As Jaron Lanier details in his interviews for the documentary *The Social Dilemma*, Facebook and Google can essentially promise advertisers specific behaviors and get paid based on certain conversions. So if their revenue is dependent on users making those "choices," it is in the platforms' interest to design systems that maximize the conversions regardless of the impact on us as individuals or as a society.

Levitin speaks to our inability to make good, rational choices when overwhelmed with information, and in the twenty-first century, we live more or less in a perpetual state of information overload. Constantly presented with more information than our brains can process, we inhabit a scenario that promotes impulsive decision making and creates more revenue from online advertising and online commerce.

What we have is the exponential acceleration of a common swap at the heart of American capitalism: accomplishment for meaning. But this time, the media industry has sped up the process and turned it on us in order to build inventory by monetizing our anxiety and fear. These days, we look back on the excesses, cruelties, and waste of the nineteenth- and twentieth-century Industrial Revolution with amazement at what we considered acceptable, what we considered normal. A decade from now, how will we look back on the treachery and waste—the exploitation, the loss of choice and privacy, the misinformation and disinformation—that we tolerated during the Information Revolution?

## PRIVATIZATION OF PUBLIC GOODS

As if obfuscating their practices of exploitation weren't alarming enough, the private platforms that form the foundation of the graph have also taken the reins on fundamental public conversations about privacy, security, and shaping the incentives of storytelling. When public institutions that didn't understand the questions at play (or didn't recognize the importance of their leadership) abdicated this responsibility, we permitted media companies to act as the locus of these essential public goods in our modern society. So beyond just handing the henhouse keys to the foxes, we've also asked the foxes to define the rules *and* guard the door.

Mark Zuckerberg has routinely and famously recognized that Facebook is more like a nation than a company. And his platform, in its ubiquity, has set the norms and standards of governance for how people control, understand, and trade their privacy for access in the graph. The boundaries of the right to privacy remain an open debate in the United States, continually being hammered out in the courts (and in the court of public opinion). Yet the online standard is now clear: our information is public by default.

Most people do not recognize how much of their lives are exposed—and indeed, for sale—in datasets as far-reaching as the location history of their phones (sold to the highest bidder by mobile telecom providers) and their internet browsing histories (sold by ISPs). Many Americans are hung up on the possibility of government surveillance, but our private information is already an asset of the very companies we rely on for access to information. We live in an era when our safety is for sale.

Our history is another example of a public good that has been put up for sale. And in addition to the blurring of lines between information and entertainment, between news and opinion, between analysis and fact, we are also witnessing the convergence of history and entertainment. History masquerading as entertainment leverages the same sensationalism that news sources use to compete for attention, along with reinforcement of the status quo and celebration of outrage and conflict. Let us not forget: making history entertaining is different from commercializing history.

Disney now controls its own canon alongside the Marvel universe and Star Wars saga and effectively owns most of modern, mainstream mythology. While some efforts have been made to expand that canon and step outside the reinforcing hegemony (Captain Marvel and Black Panther especially come to mind), commercialization generally precedes a loss of principle. As companies like Disney hurtle toward more homogenous narratives—less connection to tradition, less nuance, less spirituality, less recognition of the value of wisdom and the past—non-hegemonic (namely, non-white) audiences are driven to the edges. These groups participate less and less in a shared cultural narrative. Just as the American Dream no longer represents a unified story for our nation, our mythology no longer represents a unified code of values for our history.

The internet's original creators built certain key technical concepts into its underpinnings that forced a certain kind of openness, as Jonathan Zittrain outlines extensively in his 2008 book *The Future of the Internet— And How to Stop It*. But because they were reliant on public infrastructure at the time, they never envisioned the kinds of commercial incentives that drive how the graph functions now. The original networks were built with and on public resources by DARPA and the academic community. But the internet is no longer public. The principles for governance that might have steered us toward user-centric, community-centric, and democracy-centric experiences never received the kind of public airing needed to embed them in the regulatory frameworks that guide the evolution and expansion of technology. But they could receive such treatment. We can pick up this conversation anytime we wish. And we should.

Our mythology helps us understand the world. It offers us both shared values (our cultural definition of "hero," for example) and a common metaphor in the form of archetypes and communal stories. Control over that mythology represents an immense power to control society itself. The potential for that control—for greater connectivity to drive greater participation, for exposure to more perspectives to drive greater inclusion, for easier access to information to reduce barriers to entry and drive a richer civic life—is alive in the graph right now. But as long as we allow our public needs to be subsumed by private, commercial incentives, that potential will go largely unrealized.

## WELL, SOMETIMES IT *IS* THE CRIME

Much of the economic pressure undermining our capacity for public discourse, for leadership, and for society to take advantage of (and live up to) the potential of modern media is fundamental to the individual business models in this industry. There are also notable economic consequences attributable to the opacity inherent in the attention economy. But these problems within the media and technology industries have been intensified by the same broader economic trends in America over the past century—especially the gradual relaxation of antitrust enforcement and the sanctioning of highly monopolistic markets. The argument that Facebook is too big not to break up because of how much the company dominates the market for social media is the wrong lens for viewing its antitrust abuses.

It is not the market for social media that technology companies are abusing, but the market for digital advertising. With close to 70 percent of that market, Facebook and Google dominate as a veritable duopoly. With so much of our consumption happening in digital spaces, anyone interested in driving a broad cultural conversation must go through them. Much of our information discovery has become dependent not on organic search and serendipity, but on paid access to the attention inventory these two companies control. Even access to the organic communities that we have invested time and care to build often requires paid content promotion: Facebook manipulates what is seen in every feed on its platform, including inside the Groups we create. And the US Department of Justice has finally begun to pursue Google not just for its dominance of search advertising, but for its use of that dominance to exercise monopoly power over other product markets.

Monopoly power in this sector has the same basic consequences for our economy as all other monopolies: the lack of competition that protects profits also dramatically suppresses innovation. This loss of innovation has especially profound consequences for media and information because it reinforces the belief that only ad-based media can survive. Without more innovation, the barriers to experimentation with new models continue to rise until they can work only at massive scale. Small innovators are crushed, copied, or acquired by one of the two dominant platforms and never manage to prove the viability of their models.

Beyond these economic consequences, the control of information gives these companies disproportionate power compared to dominant actors of previous eras in terms of their ability to control culture. And their failure to design for discourse in a society reliant on digital experiences for community underscores the enormous responsibility they wield. The tyranny of outrage, the impetus toward polarization, the lack of choice, and the destabilization of effective leadership—all are exacerbated by these monopolies. These civic consequences should trouble regulators and citizens alike. We should be more demanding of new designs and tighter consequences for these companies.

The new platforms and technologies may dominate the most recent "too big" conversation, but traditional media consolidation and lack of diverse ownership at a local level have profoundly narrowed and nationalized local news and local media as well. In 1983, 90 percent of US media was controlled by fifty companies; by 2017, that number was down to five. And nearly sixty years after the Kerner Report cataloged how the lack of diversity in media ownership was contributing to racial tension in America, Arie Beresteanu and Paul Ellickson's Duke University study in 2007 found that media ownership is more than 90 percent white—and more than 70 percent male.

Much of our conversation about diversity in media centers on representation and diversity of voice, both of which are essential. But lack of diversity of all types at the ownership level is equally problematic. The Federal Communications Commission (FCC) has failed to enforce media ownership rules consistently, much less understand the consequences of rules that have not been modernized to acknowledge the relationship between channels as edges and the increasingly limited range of information in small markets. This failure is largely to blame for the fact that too many Americans live in an information desert.

Recognizing that Facebook and Google have created opportunity for themselves is essential—and this is exactly as it should be. They are not public entities. They are not social impact organizations or charities. They are *private* companies optimizing their business for *private* profit. They represent the exact moral limits of markets that Harvard political philosophy professor Michael Sandel details in *What Money Can't Buy*, his exploration of where and how we should expect private markets to make ethical choices and provide public goods. And they have become engines with the potential power to sustain the cultural hegemony of the United States as we face the third wave of global empire, this time in the form of information warfare and control. But missing from their role over the past decade is a *public* conversation about the principles our society wants to see at work—the values governing our economy, our storytelling, and our social concepts like privacy, all of which are adjacent and interconnected with these modern media platforms. That vacuum has been filled by the platform companies themselves, which have taken over the architecture of storytelling and now wield unprecedented power over our information landscape, largely in algorithmically hidden secrecy.

As a society, we have by and large acquiesced to the visible, known trade-off—we give up our choice and access to our data in exchange for information and perceived social connection—in the form of advertising. This comes at the expense of an invisible, unrecognized trade-off: the loss of a public sphere intended to foster the common good.

# 4

## YOU MAY ASK YOURSELF, "HOW DID I GET HERE?"

AFTER ENDURING ALL the history (which has offered helpful context, I hope), math (sorry about that), and economics (also, sorry), you may be asking: What does the evolution of media and information really mean for our civic life? What does it mean for us as individuals and for our leaders?

At a fundamental level, we rely on modern media for the information necessary to be good citizens. Our leaders rely on and leverage these same systems in order to lead. The misplaced incentives and outdated business models that affect our information habits and our relationships with one another affect our leaders just as deeply. Leaders are human too, striving to balance their desire to serve with the pressures and incentives of systems that push them toward unhealthy behaviors.

The graph, and our inability to adjust appropriately to it, has led to isolation, uncertainty, and a pervasive (if often hidden) lack of agency. Exacerbated by the economics, this isolation and the ensuing confirmation bias have been weaponized against everyone, including our leaders. Striving for greater connectivity, understanding, empathy, and productive debate—the great promises of our modern media platforms—is like swimming against that multibillion-dollar stream. The consequences are even more damaging when we see how this new landscape undermines leadership, democracy, and our civic institutions. The more our choices are made for us by systems designed to exploit us for commercial purposes, the less individual agency we have, and the less we live in a system of self-government.

## THE GRAPH

The graph is a conceptual and mathematical model that helps explain the information landscape, yet it is also a physical map. Plotting nodes with a greater number of shared edges more closely reveals clusters of people who share relationships and also, in most cases, share an information diet. Proximity is the measure of how close we are to other nodes—how many relationships to other nodes and edges we share. The power of proximity to drive behavior and increase density is at the heart of the "network effects" that start-ups love. If we rely on proximity to dictate what we consume, we end up with more pressure toward like-mindedness.

The cognitive and psychological power of confirmation bias is profound and not at all new. No one likes to be confronted with the limitations of their own thinking or admit we have an inaccurate understanding or harbor implicit biases. That's why we are all likely to appreciate and value whatever reinforces our opinions, whether people or content. But when we lack mechanisms or processes for identifying what is authoritative and instead substitute proximity, we unconsciously submit to confirmation bias. What's more, we unintentionally subvert our discernment to others' judgment because what is being consumed and shared by people close to us drives a major proportion of the voices and content in our view of the graph. In other words, we don't choose what our view from the graph looks like—our close connections choose for us. So our own choices often are not only obscured by algorithms, but also replaced by the choices made by others.

In popular culture, this trend frequently makes for homogenous and boring experiences. For civic life, however, it has even more significant repercussions: on the physical graph, we see more cognitive distance between us and those with other opinions. This distance creates more space for leaders who are interested in the power of "othering." When faithless leaders consolidate power within homogeneous segments of the graph, they can insinuate blame and pain into those spaces.

And that's exactly what we see happening today.

## PREDICTABILITY VS. OPPORTUNITY

In a world where everyone is forced into the posture of participant and where power may or may not flow from institutional perches, all communication has become more multivariate. Leadership necessarily takes on new tones and requires new skills.

As individuals, we have generally found that being released from the role of passive audience is both freeing and disorienting. Ultimately, it offers greater opportunity to play more active and more differentiated roles in society. For candidates and leaders, the reality that all marketing is community organizing might seem like an obvious opportunity, but the loss of predictable, stable audiences who are easy to reach is just as disruptive to their communication as it is to a brand. The conventional wisdom of campaigning (just like the conventional wisdom of advertising) is under new pressure from these new systems. Predictable audiences were relatively easy to address and with stable, channel-based communications, estimating the effectiveness of this messaging was simpler (if inaccurate).

The unpredictability of the graph creates opportunities for new types of leaders to emerge from less predictable places. Just as traditional gatekeepers have lost control over information and attention, the traditional political power brokers now have less control over who ascends to leadership. And cultural relevance can be translated into political power more easily when the barriers to cultural participation are lower. The markers of authority and credibility that our old gatekeepers often used were deeply flawed: elitist, racist, and sexist. Their loss of control over what a "good" leader looks like is a massive step forward. Our representative democracy can become more representative, and we enjoy an increased capacity to question an aging status quo that never served all of us. Our inability to control the information landscape also diminishes our ability to control campaigning, which can lead to good outcomes for new leaders emerging from unexpected places (Representative Alexandria Ocasio-Cortez), but can foster dangerous outcomes from new leaders emerging from culture without credibility (President Donald Trump).

## AUTHORITY VS. POPULARITY

As the control of gatekeepers has collapsed, we have seen the rise of new and more diverse voices in leadership over the past decade, which is unquestionably good for democracy. Citizens seeing themselves in leadership is fundamental to that leadership feeling and being representative. However, without an effective substitute for the authority function that gatekeepers used to provide, we struggle to identify who is authoritative. Instead we rely constantly on other substitutes, each with its own problems for leadership.

Authority is perceived online using one or more of three major substitutes: volume, proximity, or visibility. But none of these guarantees us that the idea or person held up as an authority actually is one, and each has its own problematic effects on our ability to participate authentically.

If we rely on volume for authority, we may end up confusing it with celebrity. This culture of celebrity is often viewed as one of the central consequences of the attention economy, but it is less about the basic economics of attention and more a consequence of our misinterpretation of attention. When we interpret attention as authority rather than simply as popularity, we begin to unbalance our understanding of the world and the narrative landscape. Volume and reach do not mean authority or credibility or expertise or value. Often (especially as organic reach on platforms like Facebook has diminished), volume may not even mean popularity; it may just be a function of who is willing to pay the most.

For leaders, this celebrity misinterpretation produces another "if you can't beat 'em, join 'em" mentality that drives a focus on popularity rather than authority. It adds enormous pressure to compete on volume by spending resources on paid media. Interpreting paid media as the path to balancing volume can also produce inaccurate assessments of the power of personal conversation among groups *outside* campaign control. These conversations, especially inside Facebook, are notoriously hard to quantify because much of our personal engagement is not visible to external organizations. While Facebook has intentionally reduced organic reach for organizations in order to drive more advertising dollars, organic reach between individuals remains high, especially when artificially boosted by the fake accounts and bot strategies rampant among propagandists and disinformation sources.

Related to volume, visibility is simply the ability of content to be seen, regardless of overall reach. It asks, can a story make it into our view through the algorithms that choose our content for us? Our entire worldview and our collective Overton window are subject to the algorithms that define what is visible.

Visibility defined by our streams puts massive power and responsibility for our thinking in the hands of the people we follow and the algorithms that define them. As leaders look for visibility, seeking to break through and connect with people, they can take advantage of this lack of authority to insinuate themselves into conversations where they do not belong. And when algorithms optimize for outrage and conflict, we become subject to the outrageousness necessary to be seen. We can leverage the tyranny of outrage to force our way into conversations on which we have little to nothing to offer but more outrage. And these algorithms do not reward our expertise or authority.

Because the systems we use do not recognize or optimize for genuine authority, it has regrettably become not only unnecessary but wasteful to build expertise or to focus on the nuanced, thoughtful reality that authority often demands. In the second definition of *authoritative* (who has the authority to make decisions, to set policies, to speak), we are all better served by systems that encourage us to take our seats and accept the authority of citizens in self-government. The old gatekeeper that wielded the power of granting authority also determined who had authority to speak and to act. Loosening those constraints is good for self-government. But for any future model to be truly effective, it must be possible to elevate individual authority to act, speak, and participate without undermining the importance of being an authority on a subject.

Traditionally, authority was most often a tool of the status quo and hegemony in editorial and publishing realms. New tools and greater access to distribution has helped us to loosen its grip. But without a new model that distinguishes authority, we are left to an almost impossible task of parsing the value and legitimacy of the authors producing a deluge of content all on our own. Left to our own devices, we are forced for the sake of sanity to rely on bad substitutes to make sense of the world. Those bad substitutes are skewing our understanding of and relationships to both our leaders and the most important issues society faces.

## CREDIBILITY VS. CONFIRMATION

Credibility is the quality of being worthy of belief. It attaches to the content itself and, much like authority, suffers from the lack of clear identification and from ineffective substitutes. In a real-life stream, each molecule of water is indistinguishable from the next. As more and more of our content consumption moves toward online streams and feeds, the platforms benefit more—in terms of maximized attention and continuous engagement—from the indistinguishability of credible versus incredible content than from accuracy. Power-hungry leaders win an advantage when they are willing to say whatever might be in their minds declaratively enough to be believed because we don't have a good mechanism for recognizing when the content they create or share is not credible. This indistinguishability enables politicians to distract from meaningful discourse and dictate the frame of conversation regardless of credibility.

We have become so attuned to substituting credibility with confirmation bias or with proximity that we now find it difficult to distinguish when leaders we *like* might be misleading us (intentionally or not) or when leaders we *dislike* might be opening our eyes to something important. Authority and credibility are key components of building public trust. When substitutions lead us astray relative to information, they also recast trust as agreement. For trustworthy leaders who are not trying to exploit a lack of credibility, we lack strong reinforcing incentives that can do the work of creating credible information.

Being an authority and focusing on credible information ought to be fundamental to building trust in both leadership and institutions. But when neither of those concepts works as needed, trust also fails. In the past decade, we have seen that breakdown worsen in favor of familiarity and, ultimately, partisanship.

In addition to crumbling trust, the failure of our systems to help us understand and distinguish credible information also results in a decrease in the value of expertise that society needs to function. Undervalued expertise is especially problematic in a crisis that requires technical knowledge—case in point, public health information during a pandemic. Just as we substitute proximity for other types of authority, we begin to rely on proximity for validation of expertise on technical matters as well.

Technical expertise is essential for us to make healthy decisions, and yet it often comes from voices and institutions that are the least well suited to the new landscape of storytelling. This sets up a tricky dynamic for society where even (or especially) in moments of crisis, the information most needed is least surfaced. Unsure about credibility, we become dependent on close relationships or celebrities for technical truth and expertise—a confirmation bias that is decidedly dangerous and leaves us susceptible to manipulation.

## DIRECTED VS. UNDIRECTED

Just as unpredictability offers new opportunity, for leaders who want to listen, connect, and be in genuine dialogue with the people, undirected edges magnify their ability to maintain connection and to listen at scale to constituents. Modern media systems are chaotic, always-on conversations—a wildly different dynamic from the stable one-way message delivery of broadcasting via directed channels. What this requires of effective leaders is a willingness to stay in conversation until their communities have heard what they need to hear, said what they need to say, and feel heard.

This behavior is fundamentally different from the actions of most traditional leaders who want to be in conversation only if they can speak unchallenged by uncomfortable questions. The mechanisms and organizational dynamics of listening effectively at scale and leveraging this potential lag in power leave many campaigns struggling for the tools and skills needed to get beyond surface-level community engagement. The proliferation of undirected edges also demands an understanding of the extreme portability of content.

Politicians are used to being able to comfortably say different things in different rooms to different communities. But we are now in a world where any of that content might find its way to unintended audiences. In 2012, Governor Mitt Romney was speaking to a private fundraiser about the state of the presidential race when he said that 47 percent of voters don't matter. What he meant was that 47 percent of voters had made up their minds and therefore their opposition was locked in. But saying that a large block of Americans don't matter came across as offensive and out of touch. This posture was problematic for sure, but when such a statement is traditionally confined to a small private conversation, it is not debilitating for the campaign. This time, however, the conversation was filmed by a waiter at the event and shared with *Mother Jones*. The Obama campaign ensured that every possible undecided American might see it, by sharing the quote on social media and inserting it in television ads.

This cautionary tale is not about the danger of being taken out of context (a real but separate issue). The lesson here is that in an extraordinarily interconnected, unpredictable world where conversation across media platforms is the norm, we must assume that all conversations are ultimately connected. For leaders who are more interested in command-and-control, top-down machine politics, undirected edges are to be avoided and feared.

## POWER VS. ATTRIBUTION

The graph expands the power of who gets to tell stories. For leaders, this reorganization leads to a fundamental shift in how power is accrued and expressed, with two distinct consequences. First, power does not necessarily or stably attach to institutions. Second, with more volume of content and more complexity of distribution, the power of any one node or of any one piece of content may decrease *or* increase in ways that are hard to predict, ways that depend on factors largely beyond our control. And third, the attribution of what outcomes the expression of that power creates is much more complex.

In the same way gatekeepers of old restricted access but maintained collective authority and credibility filters, the addition of more voices to our public sphere means that power is widely available (not equally distributed) throughout the graph and that institutions are participants in an enormous dialogue rather than determinants of it. Institutional power used to be guaranteed by access and control. Now institutional power is determined by the same capacity for captivation, relevance, and ability to leverage the incentives of the graph as any other node. Some institutions retain substantial power because they remain effective storytellers in our new world. Some retain power because they retain control over certain types of content (like CNN's broadcast channel) or specific edges (like Facebook). But most public institutions (e.g., the local school board, the White House, the National Institutes of Health) find their power reduced significantly by a lack of skill in leveraging the world as it has become.

For leaders, this realignment requires a shift in perspective about what it means to be relevant and a humility about what it means to express power. Up to now, leaders have been able to comfortably assume that the political power inherent in their position translates automatically into storytelling power. In fact, storytelling power and its ability to drive culture and behavior often precede and even supersede political power. We have become so attuned to cultural power that we often confuse it with institutional or political power; our attention-obsessed, celebrity-mad media systems mislead us about where different types of power truly reside. Community leaders who master our new graph-based storytelling can rise to authority rapidly and express profound influence over their communities. Elected officials who master storytelling (and they are few) multiply their institutional power exponentially in combination with cultural influence.

Donald Trump is an example of someone who was able to translate mastery of media systems into cultural power into institutional power as he reset the terms of the 2016 GOP presidential primary. The traditional leaders he competed against in that primary, mostly very comfortable with existing power structures and status, simply looked on agog at a phenomenon they did not understand.

Representative Alexandria Ocasio-Cortez is perhaps an even better example of someone able to build power via unconventional storytelling and then translate that into institutional and political power, all while maintaining her handle on cultural influence. Her unwillingness to conform to traditional standards of communication once in office is evidence that her ability to maintain direct connection and authenticity among her community is fundamentally different from what we typically see in our elected officials. And her refusal to accept various standards and kowtow to that status quo riles and confounds Democratic leadership nearly as often as President Trump does Republicans.

As power shifts between nodes, and as new types of content emerge as powerful, our understanding of what campaigning looks like and what kind of campaigning is most effective is also in desperate need of revision. Political campaigns' reliance on our old understanding of power means they misattribute the effectiveness of old methods and undervalue new ones in ways that lead to poorly designed campaigns and unexpected outcomes. Leaders who remain convinced of their institutional power, independent of their mastery of (or failure to master) the new architecture of storytelling, are often surprised by the loss of political power that accompanies their loss of relevance and cultural power.

This redistribution of power and the recalculation of how power works offer a critical opportunity for communities and organizations that traditionally have been disempowered by institutional control or locked out of systems by gatekeepers maintaining an (often unjust) status quo. For those unserved by the status quo, this shift is an opportunity. For those in power or more comfortable with traditional expressions of institutional power, it's a nightmare.

## COMMUNITY VS. AUDIENCE

Like unpredictability and the rise of undirected edges, the shift from users as passive audiences to participatory communities is both a massive opportunity and a positive pressure point on the basic engine of self-government. But that's only true if the participation is used for public good and leveraged by a certain (unfortunately uncommon) type of citizen-centric, servant leadership.

As leaders have become more disconnected from the people they represent and more focused on the ideological edges of politics, community has often become a mechanism of opposition. Grassroots engagement is a lever of resistance to entrenched power. The opportunity of embracing the broader distribution of power in the graph is that the potential of aggregate power across a large, diverse, active community wildly outstrips any individual potential power—no matter how many direct connections a leader might amass or how much charisma that leader might have.

The key to leveraging community power is confidence in the ability of shared values articulated via compelling stories to drive coordinated, cooperative action and to inspire (not direct) behavior. Inspiring people and drawing them into shared effort can expand the power of leadership in communities that are heavily connected to each other—not just to the leader as a central common node. These connections are more resilient and possess far greater capacity to diffuse content through the entire graph rather than just the dense and narrow (even if large) community of first-order connection.

TOGETHER, THE TENSIONS between the consequences of modern media's evolution to the graph translate into an architecture that supports more community-driven politics. The fundamental shift encourages emergent and generative ideas about leadership while simultaneously undermining traditional power structures in ways that can drive more power toward individuals. Like the cyberutopian visions we were promised in the early days of social media, the positive outcomes of these shifts are not automatic: they depend on our collective intentions, incentives, and actions.

As institutional nodes attune to these changes and embrace the new architecture, they gain the capacity to leverage their outsized infrastructure and higher degree to great advantage. The support of individual power in the early days of this evolution to the graph has been largely due to timing and to who has been able to adapt quickest, not to individuals inherently or permanently being able to leverage the new landscape.

The sum of these implications for leadership is a civic life disrupted. All the intended and unintended consequences of the rise of the graph—ineffective leadership, democracy in turmoil, citizenship questioned—can be summarized by three fundamental failures that we must address directly and forcefully if we hope to reclaim our civic life.

## FAILURE OF CHOICE

For individuals, less visibility and less control over our own information choices point toward a moment of maximum manipulability. Hyperpersonalization and obscure incentives are the bugs that propagandists and bad actors use to control people who are isolated from and unaware of the choices being made for them. The rise of state-based propaganda from Russia influencing American culture and elections in the form of disinformation campaigns is one example of the danger of large-scale, institutional bad actors who can leverage this landscape; the shifting control of cultural conversation by the Trump administration via misinformation is another.

Our lack of regulation and control mechanisms for credibility and authority present colossal problems at the intersection of greater opportunity and breakdown of social cohesion. In a democratic system, false choices or outright control of choice limits the most fundamental right of self-government. Yet we have created an entire architecture of media and information based on *apparent* choice that drastically limits *actual* choice.

When we cannot control what we see, we cannot control how we understand the world. If our worldview is controlled by others, ultimately our behavior is in their hands as well. Our political choices become automatic, predetermined, and manipulable by those who control the levers of power or can buy access to them. Controlling the stories that we collectively hold as culture means controlling society, and that means we are living not in a system of self-government, but in a system of social control masquerading as self-government, of apparent choice masquerading as actual choice. And in a world without actual choice, new leadership emerges rarely and only as an outlier.

Conventional wisdom and the status quo that governs much of political tradition have always worked to reinforce leadership that looks and behaves a certain way. In the messy graph where we now live, the elevation and leadership of new and expected voices is the norm. Those voices finding their way into seats of power or opportunities of service is still too rare.

## FAILURE OF LEADERSHIP

Without some kind of shared narrative about who we are and how our institutions are meant to function, it has become impossible for even the best elected officials to lead effectively. As individuals, we are not incentivized to converse or debate with others, and neither are our ostensibly representative leaders. They live in the same narrower and narrower ideological environments as we do, driven by the same algorithms that cut us off from the world, and are just as susceptible to misinformation and manipulation. But our leaders are also subject to economic and political incentives that dramatically alter their behavior, narrow the ideology of our political parties, and frame the attention of our political leadership. And in an increasingly corrupt environment, fewer and fewer of our public servants are even trying to push back against these incentives.

The graph we inhabit and the business models that reward our individual extremism also reward ideological extremism from leaders. Platforms that encourage outrage reward outrageous behavior; as that outrage fuels our ever-increasing percentage of political communication and increasing numbers of average, nonactivist Americans pull away from political conversation, our public discourse is increasingly dominated by these extremes. We suffer from the tyranny of the loudest, while the 70 percent of Americans in the Hidden Tribes Exhausted Majority push back from the table, feeling utterly unrepresented by either party and exhausted by all the shouting.

Then there are the economic and political incentives codified in a campaign finance system and a primary voting process that, in most states, push would-be representatives toward the ideological edges of our two-party duopoly. Our electoral systems magnify the extremism introduced and incentivized by our media systems, including a news media business model that demands a focus on conflict. Political coverage in the news has narrowed to the competitive horse race of elections rather than encompassing the vast and more essential majority of democratic processes: the actual governing.

Our politics have become entirely dominated by increasingly extreme popularity contests over who is in charge and are not about governing or leadership at all. Our selection process for who will lead us has always focused on the wrong skills, but increasingly the process is optimized to identify people who are the best at campaigning, not the best at governing. The skills of governing are nearly absent from our political process and practically invisible from our culture. Leaders eager to lead and to focus on service are often drowned out by leaders focused on power. Those in power are far too willing to ignore the responsibility of leadership, sustaining focus on the conflicts that allow them to maintain power rather than recognizing that in a democracy, power is borrowed and meant to be used in service of those represented.

The only incentives that drive leaders toward justice, nuance, moderation, and real representation are moral ones represented (however imperfectly) in our founding stories and in the narratives we pull from our own communities. Without those shared stories, even the moral incentives start to collapse in favor of short-term, power-seeking behavior, with each ideological edge along the spectrum taking turns trying to impose its will on and extinguish the other.

If we expect our leaders to represent everyone they are supposed to represent (not just those who voted for them), then we need systems that drive them toward and then reward them for listening and leading broadly rather than nurturing extremism. Democracy is not about elections. Democracy is about self-government, about imbuing the people with the power to lead. But as more and more people disengage, democracy becomes captive to those who remain: the power-hungry, the corrupt, the corporate interests willing to exploit public systems for profit. Dynamics that reinforce narrow perspectives, that incentivize the fracturing of shared stories, make it impossible to lead. There are committed, dedicated public servants at all levels of government. But just as reporters are captive to a news business model that works against their best moral instincts, so too are our best leaders trapped in systems that only reward the worst behaviors. Too many of our leaders work for us only opportunistically or tangentially. In the same way a news organization optimizing for attention will only accidentally engage in the truth, so a democracy optimized for conflict and corruption will only accidentally engage in equity and justice—or when forced by circumstance and crisis.

The decentralization of media and the resulting greater connectivity was supposed to yield greater transparency, accountability, and access to government for people. But our internet is private: run by private companies, incentivized to private gains. We have no public internet optimized for public discourse, public good, or public governance. And with journalism, out of desperation for survival, opting for the same perverse incentives, our chances of effective self-government are slipping away in the face of exploding activism and declining overall civic engagement, attention, knowledge, and participation. We are losing our grip on the shared narratives that provide the fabric of moral leadership.

## FAILURE OF GOVERNANCE

Democracy is a system of faith. All these negative factors surrounding our political and social climate lead to a pervasive cynicism about politics and government that is anathema to democracy. We are living through an era when civic life feels not only exhausting but pointless. We often wonder whether our votes and our voices matter. Democrats lament that only about one-third of the country voted for President Trump in 2016—but only one-third of the country voted for President Obama in 2012. We have not created a system or habits of participation that ensure strong, broad majorities.

Our disconnect from national leadership pales in comparison to our near-complete disengagement from local community leadership. Practically no one goes to community meetings, runs for school board, or participates in town leadership, where a huge proportion of the actual governing still takes place. As the media pulls our attention toward state, federal, and national contests that all sound like nationalized conversations and feel ever more remote from our daily lives, our civic life is bankrupted by a lack of agency, a lack of energy, and ultimately, a lack of participation. Our self-government gives way to governance by a largely corrupt political class we don't know and who do not know us, and our indifference to them is a descending spiral of cynicism and disengagement.

To be clear, this is not about apathy. We do not care less about our communities—we simply see less point in participating in a process that does not seem to care about us. Blaming voters for the failure of leaders and systems is not only unproductive but offensive.

The language we use in the stories we tell about how our governance is failing us is also contributing to the downward spiral. Games that are "rigged" are not worth playing: they do not reward participation. Systems that are "broken" are not fixable. Whether it's President Trump talking about rigged elections or Senator Bernie Sanders decrying a rigged economy, this type of language undermines our ability to confront and restore them.

The systems of governance in a broad sense—media, community, institutional, regulatory—are neither rigged nor broken. They are working exactly as they were designed to work by imperfect (or at times, outright malevolent) people. To bring back effective governance, our systems must be redesigned, and that redesign process must include more of us and more of our participation. Otherwise, we risk simply reimplementing the same bankrupt thinking that has put us in this downward spiral.

How do we begin? By reclaiming participation. By demanding that our institutions target the public good. By discovering and investing in mechanisms that support our pushback against the multibillion-dollar stream. By reclaiming our attention from useless stories that are easy to monetize by our existing architecture. And by returning that attention to the things that matter in civic life.

**DEMOCRACY IS NOT** incompatible with the graph. Our civic life hasn't been irreparably destroyed by social media. However, there is no going back to the old models as a solution to this upheaval. We must embrace the tools and the architecture on which our new systems are built. We have to recognize the opportunities presented but not guaranteed by them. We need to evolve our regulations and institutions to suit how we want media systems and storytelling to drive society and civic life. We must create new models with incentives that align with what we really want for ourselves and our shared story. We cannot cede that power and responsibility to private companies whose incentives may or may not be for the public good.

There is a future available to us where our civic lives are vibrant and interconnected. Where individual power drives institutional power. Where greater access and agency result in more genuine representation. And where self-government is enabled and enlivened by greater connectivity. It is a future just beyond our present, and it requires us to reengage and redesign with intention the systems as they are—not as we wish they might be—so we can make them what we need.

# WHERE DO WE
# GO FROM HERE?

I F ALL THESE consequences—uniqueness in the graph, the indistinguishability of content, the tyranny of outrage, our loss of choice, the failure of leadership—are the result of moving storytelling from channels to the graph, is our only hope to unwind that evolution? If our predicament stems from our failure to understand this new architecture, can we just go back to a simpler information environment? Would going back remove the obstacles that prevent a healthy civic life?

The answer, for two important reasons, is no.

*First, we value greater freedom.* The system of channels and gatekeepers was easier to understand but less free. Our roles, our choices, and our voices were far more restricted. Pining for the "good old days" of a single source of national information, like Walter Cronkite, doesn't serve us as individuals or as a nation. It does not recognize the painful, problematic lack of inclusion in that simpler worldview. If information is restricted for the sake of simplicity, so are our perspectives and our voices. Less freedom means fewer choices. Returning to an era when a single voice reigns would require silencing others.

*Second, we value more information.* Ensuring that a more diverse set of sources has the power to create, publish, and distribute information is one aspect of our modern media landscape that *has* lived up to the cyberutopian promises of the internet's early innovators. So far, more information has mostly led to more overload—but it doesn't have to be that way.

Overload occurs when media design choices lead to never-ending streams of information, all presenting themselves as essential and demanding real-time, unceasing attention. These information streams play on our need to belong and feed our limbic responses to outrage. The cognitive load is a heavy burden; we end up having to rethink everything we see, because each node within the graph has become solely responsible for context, authority, and credibility. From the unique view of our unique node, being a good consumer of stories requires each of us to be an expert in history, media, psychology, and more. Furthermore, we have to find the fortitude to swim against that multibillion-dollar stream all day, every day. No wonder we feel either exhausted or manipulated—and often both.

But this overload is not the inevitable consequence of more information and more freedom. We should address what is not working for us but without sacrificing all the gains in content availability and diversity that have been won.

The internet promised that greater connectivity would lead to healthier democracy. If we want systems that are healthy—good for people and good for society—we need to articulate what "good" means. We need to enable each component of our media systems to respond to that guidance. And we need to recognize that this increase in freedom and information demands more from us, both as a society and as individuals, more responsibility.

Up to now in the evolution from channels to the graph, *all* the responsibility has been carried by us as individuals. Society is failing to articulate principles to guide this progression. The companies that build and manage this new architecture respond to that vacuum by chasing economic incentives rather than public good.

The new roles within our grasp demand new behaviors of all of us, from individuals and employees to executives and company leadership to government figures and agencies to society as a whole, and we all need to embrace those roles and behaviors alike. To finally fulfill the internet's early promises, we must see this for the multidimensional problem it is. Step-by-step, we must make our platforms, our voices, our institutions, and ourselves fit for a productive and healthy civic life.

1. *Redesign our platforms.* Storytelling requires structure to serve society. New designs, new interfaces, new regulations, and updated institutions can create new, effective norms and structures.

2. *Reclaim our voices.* Democracy is something we do together. Our media systems can enable us to work together to elevate, consider, and include all voices in our work of self-government.

3. *Restore our institutions.* Our institutions must ensure that the interface between leaders and citizens is balanced and productive. New models can allow for direct, productive dialogue and genuine civic discourse.

4. *Redeem ourselves.* The instinct to reject online experiences as being less valuable than "real life" is a nostalgic and misguided interpretation of humanity and society. Holding on to the hard-and-fast distinction between online and offline no longer serves us. We can embrace the power and opportunity of the graph by embracing new edges and online experiences as central to our public sphere alongside traditional offline experiences.

Even if we favored a return to the channel model, the internet bell cannot be unrung. If the promise of a functional, connected society is to be realized, our needs must become demands. They must become the explicit driving forces behind the design choices and business models used for the systems that connect us and the platforms we use to tell, distribute, and discover our stories.

Here's how we do it.

# 5

---

# REDESIGN
# OUR PLATFORMS

THE TECHNOLOGIES WE use to create, discover, distribute, and consume stories are where our personal experiences and personal stories intersect with the new architecture of the graph. The design choices that inform the interfaces of those platforms (and the incentives behind those choices) have tremendous influence over our experience, understanding, and response to the world around us. Any edge empowered by technology can be considered a platform: Google, Twitter, and Facebook, but also, say, CNN.com and the CNN television news channel. Platforms are in many ways the most tactile, explicit expression of the opportunities and incentives within the graph. Small reconsiderations and refigurings of these technologies would have profound impacts on our relationships to one another and to leadership.

The hyperbolic cyberutopianism that gilded the early development of the internet and the rise of social media—the promises of greater connectivity, of a smaller and more connected world, of greater diversity and more empathy—is, in fact, available to us. But unlike the companies that have profited off its rise would have us believe, these promises do not come with a lack of responsibility and likely do not come with the same economics the companies have leveraged up to now. The original creators of the internet built a system predicated on the assumption of public use. They created the early systems of connectivity with humility, knowing they could not predict exactly how their protocols would be used. These principles led them to default to transparency and to delay making restrictive decisions as much and as long as possible. The success of the internet as a generative platform for innovation was driven by openness and trust, and by what Jonathan Zittrain termed, in 2008, the "procrastination principle": the assumption that most problems confronting a network can be solved later or by others. Zittrain foresaw the consequences we're now facing, as these principles give way under commercial pressure to control and to grow at all costs.

Because many of these early innovators and designers assumed public use and public interest, these assumptions were taken as *implicit* truths rather than made into *explicit* design requirements. As the infrastructure and platforms we rely on shifted from public to private hands, new values began to supplant the creators' assumptions. The growing, flourishing tech companies designed interfaces to support their commercial interests and respond to the demands of high-growth business models and expected, venture-scale returns. We still rely on them to provide public goods that they are no longer designed to create. It's no wonder they aren't working well for us or for civic life: despite all the rhetoric, they were never meant to.

Facebook and Google are the starting point for a massive majority of Americans as they begin their information consumption. These two companies control almost 20 percent of all internet traffic (according to the 2019 annual Sandvine study), and they were responsible for nearly 80 percent of referral traffic to digital publishers in 2020 (according to Parse.ly's Network Referrer Dashboard). These are among the world's most valuable and profitable companies doing what they are meant to do: grow and profit. Again, these platforms are not broken or rigged. They are working exactly as they were designed for the benefit of their designers.

That we are reliant on Facebook, Google, and the like for our civic information, for our relationships, for our access to leadership, and for leadership itself is a by-product of commercial dominance. It is not an intentional design choice by society. As information has been privatized, we have allowed important public principles and assumptions to be privatized along with it and then find ourselves baffled by the results.

We should not be surprised. And when we set down our befuddlement, the path back to systems that work for us is clear: we must redesign the platforms we use for storytelling and develop interfaces that encourage discovery of a wide range of content, prioritize credible sources, and make intent clear and obvious.

## MAKE TRANSPARENCY THE DEFAULT

As we discussed in chapter three, the attention economy is nothing new. Social media did not create the challenges we face in understanding and consuming healthy information. But the obscurity of our systems, the intentional conflation of story types, and an abundance of addictive designs have made our modern storytelling systems almost unusable for public good.

Of course, not all content is about public good. Not all information is meant to make democracy better or to enable complex, nuanced sociopolitical discourse. Sometimes it's just meant to make us laugh. But since we use the same systems of storytelling for all of our stories—information, opinion, history, and entertainment—we need to ensure that those systems support all the goals we demand of them. As it turns out, our current incentives and designs are great for entertainment, and they clearly excel at outrage-driven engagement. We've got those two use cases down. What we need is a process to reconsider our other needs (especially our civic ones) and to ensure we are well served by the platforms we rely on for creation, discovery, distribution, and consumption of information.

Our content and our behavior are the lifeblood of these platforms and the companies that design, build, and operate them. We need to demand the right to see into how our participation contributes to the functioning (and revenue) of the platforms we use—and what the effects of that participation might be. This understanding represents true "informed consent," and it requires the opportunity to fully appreciate the consequences of our choices. Platforms must identify how they monetize our engagement and data in clear, human language. They must offer information meant to clarify our choice, not expedite our conversion. It matters how the platforms we rely on for storytelling make money because these companies optimize for their customers. And in most cases, as we discussed earlier, we are not customers but simply users. As this optimization continues apace, our needs are superseded by the customers' needs. For that to change, we have to make explicit our requirements for companies that use our information. We have to make our needs part of the economic transaction that drives our participation.

Without informed choices, self-government fails. Think of the relationship between a single civic story and the whole of sociopolitical conversation as a ratio; without a clear appreciation of the diversity of the denominator of opinion in modern America, our shared sense of self fails. The opacity of these choices (and the unwillingness to reveal the algorithms that shape them) is one of the most hidden, insidious, stubborn features of our current systems. At a minimum, we must reclaim the freedom to turn algorithms off, to reclaim the agency of choice—however messy and overwhelming. Choices that are made for us ought to be reviewable on demand and unmade at will.

Consider the launch of a new app or platform, like Instagram in 2010. You find your way to the new platform through a friend who tells you about this amazing new system—a new mechanism for creating, sharing, and consuming visual stories. It's a new way to interact with people via the graph. And because you're dying to understand the world around you and be connected to your friend, you are motivated to try the new platform— even if you roll your eyes at yet another photo-sharing app.

You arrive at an interface with a glittering, all-caps promise of *New and Better Content*. You see a sign-up prompt, followed by an incomprehensible wall of fine print that explains in impenetrable legalese what you're giving up and agreeing to by participating. And below that is a button begging you to confirm as fast as possible to get to the good stuff.

This design pattern represents the signup experience for nearly every app and platform built in the past decade, and it has become such a norm that we barely even recognize the trade-off as it's happening: freedom and glittering "newness" in exchange for unfettered access to—and the right to profit off—our usage, not to mention any other data the company might be able to get us to agree to provide, regardless of whether or not it pertains to our experience. Meanwhile, there is no intent to provide clarity about what we are agreeing to; in fact, obscurity is fundamental to the business model. Clarity about the company's intent to exploit us and to share the benefits of our exploitation as unevenly as it can possibly get away with would create too much friction. Openness would slow down adoption, putting a brake on the growth that the company promised its investors, and jeopardize any number of things that are essential to its commercial survival.

These companies largely hide their economic model because they are exploitative. Transparency would destroy their access to inventory (i.e., our sustained and willing usage). That their business models cannot survive sunlight is not society's fault, nor is it our problem. Despite their constant hand-wringing and whining to the contrary, innovators and entrepreneurs need constraints. Demanding that companies explicitly consider public principles and adopting regulations to ensure the expression of those principles will not curb innovation. Such constraints direct innovation toward goals that society truly wants. Regulation unleashes innovators on a common path rather than allowing them to spin off in unintended and undesirable directions, and it ensures greater awareness of unintended consequences. Guidelines ultimately enable more productive innovation that serves our explicit needs and requirements. We should not sit back and hope that our needs are met implicitly as by-products of profiteering.

Modern media companies generate most of their revenue via the indirect monetization of our attention by selling ads alongside content. Nothing new or surprising about that—it's a central model of the economics of information. To maximize our attention, these platforms opaquely change what we see by offering hyperpersonalized experiences that are designed to provide us more of what we want—and to exploit our psychology to maximize their ad inventory. The social and psychological costs are borne entirely by us as individuals, while the platforms absolve themselves of any downstream consequences for unhealthy people, unhealthy society, and failing civic life originating from their algorithms.

We saw in chapter two that these algorithms make more of our choices about what we consume than most of us realize. The uniqueness of the graph combined with algorithmic personalization means that proximity becomes a substitute for other concepts we typically rely on to judge the quality and value of information. Both "alikeness" and "nearness" are forms of proximity that get substituted for credibility, authority, and usefulness. Sometimes that is explicit in the form of friends proactively and intentionally making recommendations to us. Sometimes it is hidden in the logic of algorithms, which reveal or prioritize content that performs well with other people like us or near us in the graph.

Platform companies and media publishers like to hide behind the trope that they are "simply providing what people want," but there is a chicken-egg aspect to this that assumes people also want what they are provided. Our Facebook feeds, our Twitter streams, any recommendation engines on any website we visit, and the headlines we see on mainstream media websites are all algorithmically tested and optimized to ensure we do what the platform companies need us to do: continue to watch and to click, to engage as much and as long as possible, so their inventory only ever increases. And we don't get to know how these choices are getting made.

What logic is at work on us? How are these choices shaping our perception of the world around us? What does it mean for us to not choose certain types of content or channels? Algorithmic transparency—what choices are being made and what is being optimized for, not how the logic is built—must become a standard, required feature of algorithmic systems.

What we exclude is as important as what we include when talking about culture, worldview, and a shared public sphere. When exclusion decisions happen outside of our view, without our knowledge, and without our informed consent, we lose something essential: our agency to define our own worldview. The platforms would argue that it's less efficient for us to make repetitive choices, or that it creates too much friction between us and the content we want. But *efficient* is not inherently *good*. In reality, making these choices helps us order the world around us. Seeing our options helps us maintain contact with the complete denominator of culture, with all the stories at work—even the ones we are not choosing to consume. The truth is that these platforms aren't primarily concerned about efficiency or giving us what we want; for them, it's all about user lock-in and predictability. In the name of profit, they are robbing us of one of the most important features of a pluralistic democracy: choice.

If the monetization of attention is the primary business model for most of the storytelling platforms, the monetization of data—both about us and created by us—is a close second. Various bits of our data are used as inputs not only for the personalization algorithms but also for ad targeting that helps organizations directly choose some of the stories we consume, explicitly for their purposes. These are the revenue streams mostly hidden by the murky, impenetrable terms of service agreements we elect when we sign up for "free" services. In search of valuable data, these companies mine and leverage everything from where we are to whom we connect with to media stored on our phones to the content of our conversations. Because data storage continues to get cheaper and cheaper—and with no incentive to rein in their own appetites for ever more data that will drive ever more accurate models of us and our potential behavior—these companies store everything they can rather than what is *necessary* to provide their service.

The combination of attention and data allows these companies to control and sell our behavior for the benefit of whoever is willing to pay, from brands selling products to the government buying our compliance to foreign actors sowing discord. At a minimum, these streams should be made obvious and open. Given that we are the source of the raw materials, however, it follows that in a healthier architecture, we would also participate in the value created by the end product: we should benefit from value created via data about us or created by us.

Whether we must have rights to this data and what rights those should be remain an open debate. In the European Union, a relatively new regulatory framework called General Data Protection Regulation (GDPR) essentially presupposes that these platforms cannot be ethical and that we must be able to assert control over them at will. Central to the EU position on data protection is the right to be forgotten. In America, there is no such established right. Not only do these platforms benefit from our data while we use them, but they continue to benefit from our data after we move on. We have no right to remove data about ourselves from their systems.

Platforms that rely on our data to build more effective models of human behavior are unlikely to make it easy to remove or delete data from their systems. They are strongly incentivized not to make our data portable to other platforms or to allow for deletion (selective or complete). Tipping control of our own data to individuals, as with GDPR, may be the right balance. Perhaps when more of our public needs are included as explicit requirements in platform designs, these platforms will become better actors. Perhaps both steps need to be taken.

Private markets generally are not going to provide public goods without direction. They require direction that is both cultural (an expression of social demand) and regulatory (a set of values-based constraints intended to guide systemic behavior). When the people reclaim our power to direct society and the systems that serve us, technology and media companies will face additional design challenges in creating platforms that provide what we want and need. But these companies don't get to posture with the unbridled arrogance of being the smartest minds on the planet and at the same time claim these problems are too hard to solve.

Modern media companies will need to figure out new business models and new interfaces—guided by public values, demanded by public spaces, in the name of public good, in a system of public power and self-government. But first we must demand transparency: the power to know what choices are being made for us and to reclaim them.

## OPTIMIZE FOR INDIVIDUAL AGENCY

The algorithms of personalization are not the only mechanisms that rob us of choice. The platforms we rely on for storytelling also could make proactive design decisions that would increase our ability as individuals to exercise agency and choice about the stories we tell and consume. Transparency—making it possible to understand the choice landscape and surfacing what is at work in these systems—would be a beginning in this direction. But as platforms begin to make choices about what to include (not just what to exclude), what to encourage, and what content and behavior to promote, they must opt for design choices that enable greater control and encourage active agency among individual users.

Right now, we are unaware of what options even exist in terms of discovery. For certain types of information, we need to be exposed (whether or not we request it) to a diversity of perspectives and opinions that help us connect more broadly with our fellow citizens. Discovery relies on either intent (i.e., search or direct navigation, purchase, subscription) or serendipity (i.e., proximity or randomness). Some forms of discovery in some forums may need to be explicitly designed to complete our understanding of the denominator of American culture and society, by offering choices that intentionally fill our gaps, blind spots, and deficits, regardless of our intent or what proximity might dictate. It is a foundational requirement of healthy self-government that we are confronted with these choices and proactively exercise our power of choice: we should not be encouraged to opt out and abdicate that responsibility of citizenship.

A platform like Facebook knows exactly what we see and what we don't. It could easily surface content that doesn't confirm our thinking or that reveals new ideas. This would be a way of actively designing against filter bubbles and ideological isolation on the topics and conversations most connected to civic life, where completeness and diversity are fundamental. Rather than replacing one choice for another, it's about providing a window on the complete range of conversation—creating opportunity for people to exclude things from their view intentionally rather than implicitly believing their view is complete.

An important element of this idea of completeness has to do with the types of stories we consume. Sometimes a cigar is just a cigar; likewise, sometimes stories are just meant to make us laugh. That doesn't mean these stories carry no cultural content or narratives; they may even codify bias in ways that are controversial or outright offensive. But not all content is meant to be civic, sociopolitical discourse. Understanding the intent of a story is fundamental to our proper consumption of it. We must be given the opportunity to appropriately leverage our own agency as we parse, decode, and synthesize each message's meaning and value.

Over the past several decades, the entire concept of news has been under attack from commercialization and the increasingly partisan intent of journalism's owners. Fundamental to our modern understanding of unbiased journalism has been a regulatory framework put in place after World War II, as television became an increasingly dominant source of news and information despite the relative lack of channel and content choices. Central to that framework was an FCC rule called the "fairness doctrine," which required broadcasters to devote airtime (although not equal time) to present controversial matters and contrasting viewpoints. Over the years, the fairness doctrine was weakened due to concerns over free speech; in the 1980s, it was eliminated altogether. Content labeled "news" has become noticeably more one-sided. The news category has become a larger and larger bucket of current affairs content that includes not only information about what has occurred in our communities and the world, but also analysis of those events, commentary about the participants, partisan analysis and strategic implications of that analysis, and parody about all of the above.

News is such a large bucket now that we stretch it under the definition of "infotainment"—a slight nod, at least, toward the idea that some of what we might be consuming on 24/7 cable news is designed to entertain, not inform. The problem with this massive conflation, however, is that we no longer know what's happening in our world. Most of what we hear or see in the "news" is not just what happened, but what someone else wants us to think about what happened.

Maybe we never really had a fact-based view of the world. Maybe it's true that in the days of Walter Cronkite, we knew only what was happening relevant to middle- and upper-class white men. But in the available media of the time, there was a shared, fact-based narrative at work about events, and that narrative, however imperfect, was intentionally separated from commentary and opinion. Now commentary and opinion *are* the events, and individuals find it harder and harder to distinguish those from actual facts. It's nearly impossible to come to each piece of content with a clear context for the conversation.

This context collapse has been largely intentional, and it dramatically favors the business models of media companies. Conflation and substitution are driven by the economics and (in the case of twenty-four-hour news television) by format. These platforms need conflict to keep up their attention inventory—whether or not conflict exists. The more salacious or emergent they can make something seem, the more attention they can sustain. Staying on air around the clock means you're going to fill a majority of that time with commentary; there simply is no way to process, synthesize, and present new information while broadcasting every minute of the day.

Part of what individuals need in order to make better choices is about labeling type and intent of the stories we consume. Content labeling, regardless of the medium, would give people the chance (and the choice) to understand what they are consuming: What type of story am I watching, and what is it trying to achieve? Innovation in format may be needed as well. Say you want twenty-four-hour commentary mostly meant to titillate and confirm your existing political bias. Sure, you can have that on channel 124, labeled "useless but satisfying hyperpartisan commentary." Straightforward reporting on actual events, meanwhile, will be broadcast on channel 125 at the top of each hour for five minutes, or in real time in the event of actual breaking news—not just an all-caps chyron over the latest hot take on the same topic that was covered last hour.

The vastness and velocity of the graph mean that while we have unprecedented powers, we also have an extraordinarily difficult time discerning value. The massive delta between the power of these systems and our understanding of them represents a moment of maximum potential for the manipulation of society, and bad actors are constantly exploiting it. Propaganda is not new. State-sponsored meddling in the elections of rivals is not new. But the ease of deployment and the effectiveness of disinformation may be at an all-time high. Without new mechanisms of typing and labeling content, identifying it for what it is, we are at the mercy of those bad actors. We cannot make healthy personal or civic decisions without new methods of coding what we consume and more clarity on why we are choosing that content.

Beyond the identification of type and intent, sourcing is also essential. Word-of-mouth sharing of stories from person to person has always been at the foundation of how we learn, how we entertain, and how we inform. But our digital reality means the scale of that sharing is new—and the anonymity of much of that content creates problems. Understanding the provenance of a story and giving credit to the original storyteller or link in the storytelling chain is crucial if we are to discern and understand that story, and crucial to the continuity of wisdom. Recognizing original provenance of ideas also permits authority to accrue to authors and may allow creators to retain more economic control over their creations.

Optimizing for maximum attention inventory rather than for individual agency has led many of our interfaces to embrace continuous, infinite streams and scrolling environments that work great for mindless consumption but are not serving us well as a society. Currently, these media platforms are designed to be addictive: their business models are dependent on the experience being habit-forming, regardless of the effect of that habit on us as individuals, our psychology, or civic discourse. While there is always more to learn, it may be helpful to designate a point of completion with regard to a given topic at a given moment. We need interfaces that encourage that sense of completion, alleviate our anxiety of missing out, and help us disengage from storytelling intentionally instead of randomly. Creating new interfaces especially for the purpose of summarizing and circumscribing the information surrounding events would give us the freedom to feel informed.

The idea of setting an endpoint to productive consumption also introduces difficult questions about coverage, breadth, and exclusion (what elements are inside and outside that boundary). But in an environment where the types and intent of stories are clear—and where transparency into algorithms and choices are obvious—we might enjoy greater insight on the credibility of our sources and the value of our content.

## DESIGN FOR DISCOURSE

The digital platforms that form the graph's connective tissue function as our public sphere. And whether the platform companies want that responsibility or not, they have it. Our democracy is something we *do* together. It is an exercise in shared, collaborative power, not just an identity we claim because we host elections. That collaboration demands that we maintain a broad public conversation about who we are and what we want for our society. This conversation is what shapes our civic life; it provides the foundation for our participation, for the ideological debates necessary to define our society. If that debate is going to be complete, it needs to include all of us. If it is going to be productive, it needs to be thoughtful and healthy. But instead of hosting a complete, representative, and inclusive conversation, the graph—and the interfaces we rely on to navigate it—tends to divide us and drive us toward outrage.

The uniqueness of our view from our node in the graph is part of what gives us more power and opportunity than we had in the channel-based world. Without interfaces that intentionally support constructive public discourse, however, our uniqueness only undermines our civic life rather than enriching it.

The platforms must take steps to ensure, while continuing to advertise, that individuals benefit from both the power of uniqueness and a healthy, vibrant public discourse. First, in order to create a shared foundation for conversation, we need to be able to distinguish sources with authority and content with credibility. If platforms were willing to design systems where more credibility meant more attention, they might begin to focus on the distinguishability problem. Streams could then be redesigned to create greater potential for differentiation and greater user control over sources and types of content.

These platforms need to accept the responsibility they have acquired (somewhat unintentionally) as a consequence of dominance and take proactive steps to encourage shared experiences. Hyperpersonalized content may drive more monetizable engagement, but shared experiences can be provided alongside it. We don't have to abandon the filter bubbles we've come to enjoy (and take comfort in), but we do need to be offered a complete view of the world around us. That means bringing adjacent but invisible conversations into view in ways that make us productively uncomfortable.

One feature that helps maximize individual engagement for monetization is speed. Today's media platforms keep the flow of content just beyond our reach, so we never catch our breath. We exist on these platforms in a constant state of mild cognitive exhaustion that ensures we make instinctive decisions and impulse purchases. Our first reflexive response can be realized instantaneously without thoughtful consideration, so we can never effectively manage the overload of information the platforms need us to consume. Part of healthy discourse is nuance and thoughtfulness—taking the time to feel and think, to consider and inwardly challenge our thinking before we respond. To create space for discourse, our platform experiences must be intentionally slowed down. Simple choices—like making a like or retweet button appear only after a few seconds (or minutes) have passed since we've read a tweet or post, or adding a lag in reply interfaces to slow the velocity of comments—can give us the time we need to maintain thoughtful behavior in the engagement loop without making all conversation asynchronous.

Facebook is aware of the topics on which we're engaging, conversing, sharing content, and posting likes, and could easily maintain a parallel stream of high-quality discussion from across the ideological spectrum into our view all the time. Like the *Wall Street Journal*'s interactive digital experience that revealed the distinction between red streams and blue streams during the 2016 election, it could include summaries of engagement—how much? what type?—on a given topic. This would give us a real-time understanding of not just our own perspective, but the entire landscape on an issue. What is truly a dominant narrative? What might be missing from our perspective?

Healthy discourse is not the same as civility. Civility is generally a tool of oppression used to quell dissent and keep the status quo intact. What we need from healthy discourse is not civility, but productive conflict: a public, ongoing debate about our values and needs that respects difference, empowers individual voices, and includes the broadest possible continuum of opinion within the confines of an inclusive, productive conversation. Respecting differences means intentionally featuring diverse voices in interfaces that are diverse by both default and design, based on our individual view within the graph. Empowering a broad swath of individual voices also expands these experiences by default, ensuring that our interactions and perspectives are continuously embracing more of the world, not less.

Building and supporting this kind of discourse requires careful community management and moderation. Like society itself, modern media platforms must discourage behaviors that are disinterested in or counterproductive to the discourse and must carefully exclude intolerant perspectives that make a pluralistic society impossible. This process will require investment in human community management that will be difficult to scale, but efficiency is not the goal here; healthy civic discourse is the goal, and it is worth the cost.

The key element of this design choice is the word *healthy*. The human community managers must be trained in impartiality, negotiation, de-escalation, and deradicalization so that these explicit values are always present and intentional, even in moments of curation and exclusion and especially in moments of conflict. This exclusion is not a function of partisanship and does not mean that all healthy discourse is progressive. It means that in our society, not all opinions are acceptable.

Free expression is important, but the First Amendment does not apply to private companies. The Twitters and Facebooks of the world rely on their users to provide the lifeblood of their platforms: the content and attention inventory that they sell. It is in these companies' interest to take positions and design interfaces that maximize both. They may choose to respect the broadest possible legal view of free expression (and they generally do), but that view is rooted in economics, not ethics.

Instead, we need our platforms to openly, clearly declare their Overton window. If a platform wants to be the Wild West, it must be open and declarative about that intent—and must take more responsibility for the public space it creates and the social consequences for the community it enables. If a platform wants to curate a valid spectrum of engagement, it must be open and declarative about that intent as well. Refusing to hide the company's efforts to maximize inventory behind some thinly veiled principle is part of its responsibility as host and protector of discourse. As James Baldwin said, "We can disagree and still love each other unless your disagreement is rooted in my oppression and denial of my humanity and right to exist." Truly productive conflict requires our willingness to embrace the discomfort of challenging our ideas and sometimes our deeply held convictions, without aggressive defensiveness and without invalidating our debaters.

Interfaces within the graph should be optimized for a collective and collaborative experience, not designed for an individual and transactional one. The significance of that experience goes beyond our comfort or desires; it's about the freedom made possible by a healthy, representative civic life. That freedom requires brave spaces, not safe ones, that provide experiences that challenge and confront us with honest, productive intentions. Keeping these brave spaces safe enough to be productive is part of the equation. And while the design of our platforms is essential to that bravery, the *least* vulnerable among us—mainly, straight white men—also bear responsibility for calling out abuse and bullying, and for actively, carefully helping protect the vulnerable and for making spaces for the unheard.

In addition to promoting healthy information-seeking behaviors and content, we need help downweighting and hiding unhealthy behaviors and content from view. Everything from simple outrage to bullying to trolling to abuse can be quashed. Limiting anonymity (except in cases where safety requires it) would vastly reduce poisonous and intolerant comments and would generally increase our familiarity in online spaces just as parks and physical public spaces build familiarity in our neighborhoods. The work of Talia Stroud and Eli Pariser through the Civic Signals project has revealed how our long history with effective physical spaces can suggest better designs for digital civic spaces. They've uncovered that familiarity from regular intergroup contact—the simple recognition of others far beyond our intimate circle (or even our weakest ties)—is enough to expand our view of the world and normalize differences that might otherwise alienate us from others.

Edge cases are complicated, and platforms will need spaces dedicated to litigating complaints, reports, and open processes for contesting inclusion and exclusion decisions; these spaces should include oversight by independent boards as well as public participation from government (perhaps the FCC). Enabling and managing the conversation we need as a society may be harder than constantly stoking the runaway conversation that's destroying our civic life—but that is no excuse not to try. Creating these experiences and enabling this needed discourse requires that we have more faith in humanity. In designing for discourse, we need to trust people more, not less.

## CREATE INTENTIONAL CIVIC SPACES

Platforms that specialize in hyperpersonalization could in fact close the distance between us and our leaders, making proximity to leadership and power the norm again. When Tip O'Neill talked about all politics being local, as mentioned earlier, what he really meant was that politics is driven by what matters most to us—that all politics is *personal*. We need to build spaces intended for exactly this kind of *personal* civic engagement and discourse.

Our institutions come with powerful bully pulpits for our leaders, but not all leadership is institutional. Institutions must be updated to ensure that the interface between leader and citizen is balanced and productive, and we must encourage leadership in other venues and from other voices whenever possible.

Our leaders need specific and intentional interfaces to ensure they can effectively participate in that conversation. As leaders engage in our existing platforms, they need to be held to higher standards than other users of these systems. Leaders are not brands. Political campaigning may be merely another form of marketing or storytelling, but the importance of that engagement means it must be governed by stronger ethical safeguards. Treating our leadership with the same rules and tools as companies and brands only encourages more transactional behavior among leadership. When leaders use paid advertising tools in these platforms, standards for fact-checking and misinformation/disinformation violations should be even higher than for companies and individuals. Democracy is a system of faith, and breaking public trust leads to cynicism. Building an ethical bulwark against that cynicism should be a central responsibility of today's technology and media platforms.

As platforms mature and embrace these responsibilities, leaders must be required to engage in and with them as dedicated civic spaces. Any other extensions and modernizations of our old institutions must also require leader participation. Public engagement can't be a "nice to have" or a best-in-class feature—it must be the standard, the baseline for public leadership. Dialogue should be featured, and listening is necessary.

Distinguishing civic speech and engagement from commercial or entertainment participation is central to this civic experience. Just as distinguishing types and intent of stories enables greater individual agency and gives us back our power of choice, distinguishing civic stories enables us to clearly identify our public conversation and the stories of our leaders. The public good is not jeopardized if New Balance ignores my pleas to reinstate the particular 1600 running shoe that I loved. Whether I feel a personal connection to and genuine relationship with leadership in a system of self-government, however, is at the heart of representation and matters profoundly.

## DEVELOP PUBLIC PLATFORMS

Healthy public information spaces and civic conversation experiences are fundamental to healthy democracy—too fundamental for us not to consider them part of our basic public infrastructure. Even if private markets accept all the responsibility they can carry, we should not expect them to provide these public goods entirely. And so we have a public responsibility to encourage healthy civic discourse in the places and platforms that we already use to traverse the graph and to invest in public spaces—online and offline—that enable storytelling that elevates our civic life.

The internet and the airwaves were developed as public spaces and innovated by massive public-private partnerships that have given way to an entirely private infrastructure. We have no public internet. The pipeline is owned by massive communications companies. We access that bandwidth through cable companies or ISPs. We log on to sites and apps built by for-profit companies and hosted almost entirely by Amazon, Google, and Microsoft. Indeed, more than 80 percent of the publicly accessible cloud computing market is controlled by those three giants plus IBM and Aliyun in China (which is Alibaba's equivalent to Amazon's AWS). These companies have created incredible things: technologies and experiences that are transforming our world.

The relentless velocity of this innovation engine is largely responsible for the evolution of storytelling from channels to the graph in the first place. But just as the economics of information defines the quality and experiences of storytelling, specific incentives drive these companies' choices. Despite attempts to ensure fair access and limit corruption where purchased traffic and speed benefit only the largest companies, these efforts have stalled and been intentionally undone in recent years. While some public spaces provide free internet access in places like airports or shopping malls, few municipalities provide free access citywide. And there are almost no meaningful efforts to provide free internet access in homes outside of dense urban areas.

At the same time, we increasingly talk about access to information as fundamental to education, health, and wellness, and we recognize the digital divide as a major cause of continued income inequality and a reinforcing mechanism for long-standing structural racism in America. Unequal access to information undermines civic life by pushing some people out of the stream of public conversation and alienating them from the information they need to make good decisions. Free and fair access to physical public spaces and our right to movement, not predicated on wealth or corporate dominance, are not controversial ideas. So why should free and fair access to online public spaces be controversial?

Once we have access to information, we also need to invest in spaces and in the information itself. There is a straightforward solution to the uniqueness challenge of the graph: make a new edge available to everyone by providing free, high-quality information that demonstrates the core principles for healthier discourse. Greater connectivity does have the capacity to increase our visibility into spaces and places far from our home communities, to build familiarity with a much broader set of neighbors, and to increase the number of relationships (strong and weak) that we can maintain. A public, free shared edge could be the new public sphere that we need.

While more and more of our lives and the stories we consume exist online, the continued importance of person-to-person communication and engagement cannot be overstated. One of the truths revealed by the COVID-19 pandemic was the incredible power of online systems to entertain and support life without leaving home. A second truth revealed, however, was our incredible craving for human contact. Investment in public physical spaces for public participation must continue everywhere in the country. Green spaces, parks, and public performance infrastructure enrich our experience and extend our potential for healthy discourse.

Access and edges are the beginning, but we also need public stories. Leaders and institutions stepping into a leadership role in storytelling is key. We should also invest broadly in healthy, high-quality information. Private media and journalism remain hostage to the business of attention and the pressure of formats. Excellent, local civic information is just as necessary a public good as the infrastructure necessary to access it, and we must put meaningful public funding toward generating such information. Recent nonprofit journalism models like the *Texas Tribune*'s represent an encouraging response to the collapse of high-quality local information, but these efforts must be replicated across the country and freed from the constraints of philanthropy and grant straitjackets.

These systems—both public and private—will remain pivotal in how we tell and how we consume the stories that define our private and public lives. They offer the potential to make our information lives richer and more diverse than ever. We just need to demand that our needs are met by the tools we use and that the trade-offs made in the design of those systems are weighted toward the public good. We can be freed to embrace innovation by the certainty that these platforms are meant to serve all of us.

# 6

—

# RECLAIM OUR VOICES

OUR STORIES—INFORMATION, ENTERTAINMENT, history, and more—are our anchors to reality and the foundations of society. The stories we share with others help us make sense of the world, set expectations and standards for social interactions, and create the definitions of citizenship and democracy that establish our parameters for self-government. And right now, our stories are out of our hands and beyond our control.

Context collapse has led to our increasing confusion and uncertainty about which stories to rely on. What's more, despite more individual storytelling potential, we have become increasingly indifferent to how we use our voices and which voices to believe. But these anchors are what tether us to each other and to a coherent, governable society. Without them, our capacity to stay in community also collapses.

Without our permission, and largely without our knowledge, our information has become controlled by news media conglomerates disconnected from our communities. In their constant chase for efficiency and profit, these corporations have created news deserts all over the country, contributing to the nationalization and radicalization of American politics. Once they're published on commercial platforms whose incentives and values are generally not aligned with our own as individuals, these stories are monetized for the benefit of those platforms—not for their authors and not for us. As more grassroots voices emerge, however, these platforms need to acknowledge the people who have an intimate relationship to our social and civic challenges and empower them to construct narratives about themselves and their experiences.

The information we know as "news" plays a central, unique role in society. It is the core of how we are informed about the world and how we make decisions about current affairs. The journalism industry codified a set of principles in the 1920s that attempted to professionalize and bring credibility to the stories created by reporters; it also claimed authority for the publishers on which they still rely. The good intent and soul of journalism and journalists remains intact, but under the pressure of economics and our new architecture, the capacity of journalists to behave as the vital fourth estate (as defined a century ago) has eroded. On television, the always-on content model propels journalism away from stories meant to inform and toward stories meant to convince or frame understanding—stories powered by the same types of outrage and extremism that dominate social platforms. Much of this content is consumed and shared in countless snippets via social media. On-air talent provide biased analysis and commentary nearly ad infinitum. Providing new, useful information about what is happening in the world around us requires reporting, synthesis, and production, all of which require more time and investment per story.

A twenty-four-hour political commentary channel is not the enemy of discourse, but labeling it "news meant to inform" certainly sets up an adversarial perspective. The term *news* needs to be reclaimed so it can perform the function in society that we need to support our civic life. In order to function effectively in society, we must salvage much of the vocabulary of storytelling as part of this process of reorienting ourselves to factual information and genuine, well-intentioned voices.

We may joke about the aggregation of entertainment brands, but a small handful of massive commercial media companies (like Disney) own the storytelling that defines our modern cultural archetypes and myths. These mythologies shape our understanding of good and evil and of our roles in the world; in them, we see represented the norms and expectations regarding what is possible. Our elders used to pass along our shared history in this same form, and we saw from childhood what was possible for us in the world. Yet today's conglomerates are largely indifferent to the responsibility of carrying this cultural burden. They crowd out voices and stories that don't make blockbuster, mainstream financial sense. We are now educated by companies with neither concern for our place in the world nor recognition of their role in society.

The stories that matter most to us—the ones about ourselves, our families, and our communities—too often are told by voices that are not our own, voices too rarely under our control. We see ourselves and our communities described in the stories created by these engines of outrage and analysis and in the myths of modern entertainment, but we have no say in our own roles and are often told or shown what to believe about ourselves.

Beyond this lack of agency, we are also the content engines for the distribution platforms that drive this whole system. Facebook and Twitter do almost no storytelling of their own. In the form of our most personal stories, we provide the content that powers these juggernauts, which then leverage our voices to build wealth—not to share history, enhance empathy, or define community. They fragment and radicalize our history, trading discourse for trolling and cancellation. They leave our communities less informed, less coherent, and with less access to information and fewer tools for storytelling. And we don't even share in the wealth our stories create.

The combination of these failures represents a profound, numbing pressure on our willingness to attempt to tell our own stories. To, as Walt Whitman demanded of each of us, sound our barbaric yawp over the rooftops of the world. To venture into the public sphere and participate in the definition of who we are. And this tentativeness has a chilling effect on how representative, how valuable, and how useful civic participation is to us all. We turn away from the flame, embarrassed or fatigued by its self-importance and its irrelevance in our daily lives, and cede the public sphere to those with the power and inclination to leverage outrage for even greater power and profit.

To reverse the damage, we must reclaim these public dialogues and make explicit our public requirements. Healthy discourse in a vibrant public sphere would invite us to debate in safety, to bring our voices and take our space around a circle that is incomplete without all of our participation. And while we focus on our ability to leverage our voices, that discourse is a conversation and demands of us a willingness to listen and not just to wait for our turn to speak. Our voices as citizens are required for self-government to be just that. As Thomas Jefferson observed, when we become inattentive to public affairs, our leaders become wolves. And so our leaders also must engage in that conversation openly, with care and attention to each voice they represent. They must find not only their voices but also their ears. Our public sphere must be a discussion, a conversation that defines our America.

## DECENTRALIZE STORYTELLING THROUGH OPEN PROTOCOLS

How we tell stories has a profound impact on how they spread. From the inception of the written word five thousand years ago, and especially since the arrival of Gutenberg's printing press in the fifteenth century, the options for creating and publishing stories have been inexorably expanding. Up until very recently, the cost and complexity of these tools made access to them one of the defining barriers to entry to participating in creating culture—available only to the already wealthy and powerful. That reinforcing effect often made mass storytelling an instrument of sustained white male cultural and political hegemony. The internet promised to democratize the tools of storytelling, and the proliferation of computers and smartphones has undeniably expanded access to storytelling tools and publication. Modern media systems and tools have magnified our ability to express story and culture, extending it to a far larger portion of the population over the past two decades. Yet uneven access to computing power and the digital divide have continued to perpetuate inequalities.

The internet was founded on open protocols—technical rules and formats for how data is exchanged—and public resources. Its designers never imagined all the commercial uses that now dominate. They had no intention of envisioning all its uses: they set out to create protocols that would enable an infinite number of communications and transfer data in decentralized ways without ever having to know anything about the content itself. But because these designers assumed that users would enforce the public good, and (being mostly wealthy white men) presumed safety and universal access, they did not build protections of public need and public access into their designs. As a result, we've relied instead on regulation, governance, habits, and individual platforms to ensure these requirements—mostly to our detriment.

With open protocols, content is easily portable, and we are less likely to be locked in to any given platform. Closed protocols dramatically increase dependence on platforms and the network effects they rely on for hypergrowth. Attempts at standardization for content syndication online were effective at certain moments. The RSS (Really Simple Syndication) feed was the dominant structure for publishers to enable easy, off-site access to content and updates. In the late 1990s, Kevin Werbach, in the influential investor newsletter *Release 1.0*, predicted that syndication "would evolve into the core model for the Internet economy, allowing businesses and individuals to retain control over their online personae while enjoying the benefits of massive scale and scope." Unfortunately, lack of clarity about goals and how the standard should evolve led to a fracture in the standardization process in the early 2000s. This fracture hampered the adoption of syndication and left open the easy emergence of more proprietary, closed protocols with the emergence of social media.

RSS remained popular through the rise of early social media platforms, which soon usurped the language of "the feed" but did not permit syndication of the content on their platforms. Open, portable RSS feeds were slowly replaced by closed, proprietary social feeds on companies like Facebook and Twitter. These companies have reached ubiquitous status while slowly migrating away from (or acquiring) open protocols, adopting closed ones as a mechanism of user lock-in. Thus, we lost control of our own voices and the universal portability promised by RSS, in favor of dominant platforms more concerned with engagement occurring on their sites to maintain attention inventory.

If platforms want to focus their innovation on closed protocols, that is their prerogative. But as a counterbalance to these closed protocols, whose incentives may not align with our public needs, we must publish new standards for (and mandate the support of) open story protocols in all public platforms. The basic principles of a universal publication-syndication (pubsub) standard remain incredibly compelling. There is reason to hope for a return to that standard, or the emergence of a new model as part of the clarification of purpose and intent for the public discourse necessary to reclaim our voices.

## REVEAL CONTEXT THROUGH CLARITY OF AUTHORSHIP AND INTENT

All the way back to the seventeenth century, when John Milton (in his *Areopagitica*) reacted to the upheaval created by the first century and a half of the printing press, anonymity was regarded as problematic. Even then it was clear that only in the most extreme circumstances of safety and resistance to oppression is anonymity preferred to clear authorship. Otherwise, we must accept the responsibility of occupying the space we claim when creating and telling stories. Today, the technologies used for faking identity in content (generally grouped together under the umbrella of "deep fakes") represent a massively destabilizing pressure on our information ecology. Anonymity undermines trust and makes even effective, appropriate skepticism insufficient, driving us further toward the comforting safety of isolation and conspiracy theory.

Any discussion of authorship raises questions about the value of anonymity and the mechanisms for validating identity online. In general, our systems for securing access (*Do you have access to a certain resource?*) are more mature than those for proving identity (*Who are you?*). Clarity of authorship is essential to rewarding creators. Furthermore, it is crucial if we are to defend healthy information systems from the increasing assault of synthetic content—some meant to be more efficient, but more intended to confuse and manipulate. Investing in a trust system that can take on the responsibility of validation—similar to how we use SSL certificates with trusted authorities to manage trusted destinations—might be the simplest answer to ensuring that stories have authors who can be known and tracked.

With open protocols consistently supported across platforms, our capacity to tell stories will be less dependent on the platforms themselves; our stories and content will be more portable, and the costs of switching to new platforms (or simply opting out of ones that exploit us) will drop. But with that openness, we must begin to provide greater clarity into the creation and authorship of stories—a first step toward becoming healthy consumers of information, encouraging healthy discourse, and achieving a healthy civic life.

Required support for open protocols and open formats for content will enable greater clarity across multiple dimensions of storytelling. Such clarity is all-important if we are to accomplish more effective information consumption, increased individual agency, reduced manipulability, and healthier discourse. These protocols and formats would ease the way for standard visibility into multiple features of story, which could then be incorporated into the interfaces we use to read, watch, and listen. Unlike the opacity of the current systems, where the author, origin, and life cycle of stories are often hidden, these features would be readily exposed and easily understood, empowering each of us to make better decisions about what different stories we consume and why and to understand them in context.

Being able to readily identify *real* as opposed to *not real* is a fundamental building block of coherent discourse. Enhancing the design principles of today's media platforms around distinguishability will help us reclaim our voices and our stories, but these platforms also need to implement more effective identity validation systems and aggressive moderation to intentionally build trust.

## PRESERVE PROVENANCE BY CLARIFYING DISTRIBUTION

In a world where all content is portable, identifying the provenance of stories is nearly impossible. The prevailing discovery patterns for new information are either intent-based (search) or proximity-based (feeds). Both come with serious challenges if our goal is some version of healthy discourse. Intent-based discovery means we never discover unknown unknowns unless the search engine is explicitly designed to provide a widening frame of reference. Google, the wildly dominant search tool in the United States, is not designed this way. Proximity-based discovery, on the other hand, relies on either self-curated or network-curated feeds, which tend toward confirmation bias and filter bubbles unless we carefully and intentionally curate for diversity. Almost no one does this in practice; the algorithms designed to surface and recommend new content never do.

But what if we built into the format the concept of origination—when and where a story began? And what if we used a blockchain to track sharing, republication, and repurposing? A decentralized mechanism for tracking the history of a story would help us understand its travels and evolution over time. The life of a story could even be connected to intellectual property rights and licensing by embedding a framework like Creative Commons in the same mechanism.

The concepts of visibility and reach are fundamental to the business of media, advertising, and strategic communication, yet are largely absent from our individual consumption of stories. Understanding the overall reach of a story—and who specifically it reached—would help break us out of our small areas of the graph, freeing us to embrace a more complete understanding of what stories are dominant for us versus those that are dominant more broadly. Portable, public data on reach and consumption would also help create a mechanism for understanding "mainstream" that is not predicated on historical privilege. Giving all of society a broader understanding of what ideas are widely held and where would enable a far more nuanced conversation about American identity. It would help subdue our sense of the perceived self-evidence of our perspectives. And it would encourage the humility required to make healthier discourse the norm rather than the exception.

Such protocols can be embedded in content itself, like the EXIF metadata layer of image format standards, and can be pushed down to the packet level in order to make some of the story metadata available independent of format. In this way, news stories would have the same origin record as the photos on our phones. Expressing public standards and encouraging open protocols is key—as is requiring platforms that leverage public resources of any kind (including internet backbone, wireless services, and broadcast spectrum) to provide their support.

Combining content standards with new protocols for pub-sub architecture would marry healthier stories with healthier distribution. Making it easy and standardized to subscribe to stories and storytellers combined with the standards and protocols for content would make it trivial to also subscribe to "smart feeds" meant to surface the most credible stories or most widely read stories independent of platform or publisher. Thus, we could subscribe to the conversations and topics that matter to us, without having to wade through the mire, attempt to scale various paywalls, and reinterpret scale of communities in an attempt to cobble together a complete view of that conversation or topic. Healthy curation could be the default, not the exception. Systems meant to reveal the complete range of perspectives as smart feeds could be standard views of new platforms. And experimenting with new models of curation or new curators would be trivial.

If this pub-sub architecture also allowed for the portability of conversations in the form of comments, conversation could be aggregated independent of platform or publisher as well. This would upend the economics of attention and the need to aggregate audience at all costs; in fact, it would shift the editorial pressure away from traffic generation toward quality. If micropayments and rewards for credibility and quality were similarly embedded in the standard, we might have the beginnings of a new content-centric (rather than consumption-centric) economy of information.

It's time to reject the idea that *what we see* is *what is*. Turning discovery into a healthy process of expansion and making the expansion of perspective a cultural norm will naturally shift the tone of debate. Familiarity builds trust, which builds respect, which builds tolerance, which leads to acceptance. This chain is how pluralism reemerges as a comfortable democratic norm. And that is exactly why invalid opinion (e.g., white nationalism) must be eliminated: so it is not slowly normalized and made tacitly acceptable.

## BUILD TRUST BY REWARDING AUTHORITY AND CREDIBILITY

Without the traditional gatekeepers to maintain control (and the status quo), we have lost our hold on ensuring information comes from authoritative sources and credible content. Reclaiming control of our voices and opening up our access to storytelling requires that we now accept that responsibility. The failure to define effective, scalable solutions for determining authority and credibility is at the root of our dislocation and the breakdown of healthy discourse in the graph. New, open standards and protocols for producing and distributing content could also help us recognize authority and credibility. But to ensure that we optimize for healthy discourse, we must also prioritize and reward those goals.

It is important also to distinguish between authority (a feature of the storyteller) and credibility (a feature of the story). Otherwise, we might inadvertently create substitutes that are just as misleading as those on which we currently rely, such as popularity, volume, or some other form of proximity.

Authority, like influence, is topic specific—a quality that a storyteller builds over time on a given subject by being a source of credible stories. Building authority is often a function of acquiring and then sharing expertise on a subject. Where the danger lies is in someone who substitutes opinion for that expertise or manipulates popularity so as to appear to be an authority. But sometimes authority is merely a function of delivering on intent: you intended to entertain, and you provided entertaining content. And sometimes it is as straightforward as expressing your own lived experience. Indeed, we are authorities on our own experience and have the right to tell our own stories, from our personal history to a broader community or cultural narrative. That right does not obligate anyone else to validate our stories. Nor does it mandate that we be experts on any generalization; in fact, we must recognize our instinct to generalize our experience to all experiences and resist claiming authority we do not have.

In order to reward authority and credibility, we must create a mechanism for identifying and elevating these qualities when we see them. We must also aggregate value toward them, independent of the business economics of the tools, platforms, and publishers. A publisher or distributor may sell advertising alongside a particular piece of content, while individual readers leverage a separate mechanism to reward the creator that inspired, informed, or entertained them. It is possible for both economies to coexist, despite the protestations from publishers and platforms that without monopolizing all value created from content, they might fail. Publishers can even become aggregators of their storytellers' authority. Imagine how easy it would be to provide an aggregate credibility score for publishers, based on the credibility of the individual stories they publish, share, or distribute.

Credibility is inherently about trustworthiness. It is not a guarantee of absolute truth or utter completeness but an expression that a story is worthy of belief. In the case of stories meant to inform, this requires adherence to fact, avoidance of unlabeled commentary, and acknowledgment of bias. It's a complex nexus, but all these concepts can be identified and modeled through advances in narrative data science. Combined with learning sets and human annotation, we can build comprehensive models of credibility over time that operate independent of partisanship.

Currently, we are living in an economy of attention that feels like a middle-school popularity contest. Large enterprises have been the only ones capable of sustaining high-quality talent and high-credibility content over time, so we have been dependent on these advertising-supported publishers and aggregators to drive the content economy. But that sustainability is largely a function of averaging revenue across a massive number of bets and selling indirect attention rather than content itself. Rather than aggregating the revenue of attention, we could direct reward toward the specific stories we find rewarding, in the form of micropayments. This is not a new idea, but executing on it has required an infrastructure that would allow for consistent application of transactions and terms of transaction across platforms. Our open standards and protocols could provide that consistency.

The stories that shape society can be free, but our trusted storytellers ought to be rewarded for their contribution to our community discourse and cultural landscape. By creating an economy of storytelling where the highest-quality, most credible stories are rewarded consistently, we can give precedence to voices of true authority who actively promote our shared civic life.

## EXPAND PERSPECTIVE BY ENCOURAGING SHARED STORIES

Reclaiming our voices starts with harnessing the useful, productive power of individuals in the graph, cultivating their ability to tell their own stories and be rewarded for them. Connecting those stories to each other and ensuring that we hear new stories—and those that help us build empathy and expand our understanding of the breadth and complexity of America—is the second layer of reclaiming a collective American voice. A portion of this work relies on redesigning our platforms to enable interfaces that encourage discovery of content from less proximate areas of the graph. In addition to the platform work, we also need mechanisms for identifying and discovering bridge stories—those that help us traverse the graph in ways that inspire curiosity, not just productive confrontation.

When confronted with new information, our instinct is to identify first what is different, what is "other." This instinct is evolutionarily beneficial because in nature, "other" is more likely to mean ".unsafe." Compound that evolutionary instinct with the momentum of our current systems and the politicians who abuse them by "othering" people and opinions, however, and we have built sturdy, problematic habits about how we engage with new information and new perspectives. What we need instead is a compassionate experience that will expand our perspectives. This new approach requires content formats designed to reveal similarities between existing preferences and new ones that show the way toward what is shared.

Being constantly bombarded with countervailing opinions and stories that confront our perspective and biases, however useful and productive, would be an exhausting experience. If on top of that we were responding with as much compassion as we could muster, we would soon burn out emotionally. Ultimately, this relentless barrage would drive us back into our filter bubble for relief. Introducing into the graph a public shared edge—one that everyone is automatically connected to—would reduce some of this friction. Then, focused curation of stories that traverse that shared edge would guide us toward embracing curiosity as part of exploration and not just confrontation.

Beyond the features of individual stories, storytellers, and formats, we must ensure that everyone has access to the graph. High-quality content must not become a privilege, and the architecture of the systems we rely on for information must not privilege any one story or storyteller. A broader perspective on the regulation of media systems is part of our conversation on the restoration of institutions (in the next chapter). But ensuring that systems of distribution are neutral, indifferent, or invisible to the content they transmit is fundamental to fair access to information. A neutral distribution system would offer a fair playing field for using our rights to free expression.

*Net neutrality* has become a catch-all term for a set of reforms and regulations geared toward ensuring this fairness. The concept is essential to rebuilding trust in the infrastructure of storytelling, especially among communities traditionally underrepresented in (or outright excluded from) mainstream culture. This neutrality principle does not mean that all content is acceptable, however, or that there are no constraints on what we as a society consider valid speech. We must articulate a set of values that govern a valid Overton window for our society, to ensure that the expression of oppression and cruelty are not welcome. Whereas "safe spaces" are focused on never confronting truths that challenge our experience, these are "brave spaces": they require enough trust that we will not be attacked, so we can be brave enough to confront our own thinking openly.

Even within one's Overton window, free speech does not mean consequence-free speech. Constraints on free expression are built into the US Constitution and our culture, and they are necessary in order to remain in community with people with whom we disagree. For speech that harms others or leads to violence or other unacceptable disruptions of society, consequences should be real and clear. But we cannot smother the vital, acceptable disruptions to the status quo under the misappropriated blanket of "civility"—all too often a tool for oppression or for the restriction of speech that runs counter to or challenges white male hegemony.

The source of moral leadership is grounded in narratives that live beyond the incentives of our current systems, but we still need and crave moral leadership. That's why we need new shared stories. It is also why we must demand that our media systems acknowledge and accept their responsibility for addressing moral versus immoral behavior in public discourse and public leadership.

## BUILD PUBLIC MEMORY BY EXTENDING CONTEXT OVER TIME

Platforms designed around infinite instantaneous experiences make tracking history unappealing and close to impossible. Worse, they undermine our public memory, which keeps us from building credibility over time and holding each other accountable to the stories we tell and share. Without a shared public memory, how can we enforce consequences when citizens or leaders are lying for their own gain, or shifting positions on the same issue over time, or depending on different audiences?

Clarity of authorship will provide transparency and improve the health of our current dialogue, particularly in the moment-by-moment choices we constantly make to assess which information we should take in, believe, and share—and which we should ignore. But those choices need to aggregate over time, as do the authority and credibility enabled by each moment of transparency. A society with no history, where storytelling and information do not develop over time, becomes a society with no consequences and no uniting long-term narratives.

This lack of shared history and the resulting lack of public consequences are particularly troubling in civic life. Without public memory, how are journalists meant to help hold leaders accountable? The graph's interconnectedness has made it harder for politicians to nakedly pander to small groups, as the portability of content reveals any hypocrisy and conflicting narratives. But the lack of public memory has encouraged promises from candidates during campaigns that are later broken, with no accounting, once the governing begins. Journalists and public transparency organizations try to introduce accountability, but the unrelenting stream of new content detached from context and history makes it nearly impossible. It also encourages inflammatory hot takes from individuals who will never pay a price for being wrong.

In the same way a pub-sub architecture could help support new dynamics of discovery and reward, a system of public archives combined with embedded provenance could make this kind of accountability an automatic feature of public storytelling rather than a nearly impossible task of investigation.

## INVEST IN PUBLIC STORIES AS PUBLIC GOODS

The stories we get from and share with our leaders play a unique, fundamental role in our understanding of the world and in ensuring that our leadership is representative. Our ability to remain in conversation with our leaders is also a tool of accountability and a way to build trust. Some of the storytelling necessary to remain informed even represents an essential public good in a democracy and perhaps ought to be publicly funded and created.

Public storytelling by elected officials—narrative content that offers visibility on process, thinking, and perspective—should be a required feature of leadership. This type of information is most certainly a public good: it shapes our view of our leaders' credibility because they would be scored like any other storyteller. It also offers transparency into their strategy and behavior, as they would be held to the same open standards of intent.

Public information creation at broad scale is also a counterweight to expanding information deserts that pop up as local media outlets discover they cannot survive in a nationalized media economy that drives toward the efficiency of scale, not the value of discourse. What we are getting is the kind and quality of civic information our media systems were designed for: national, partisan, outrageous. What we need are stories that are local, fair, credible, and nuanced. Even in a world that rewards credibility, there may be no market solution to our local information needs—but in that case, we must demand that public markets treat our public stories as the public good they are.

The creation of an always-on public edge—a public option for information—would ensure the availability of such stories independent of competition with other forms of content in traditional discovery and distribution platforms and would ensure that the information necessary for citizenship is accessible and available to everyone at all times. It also would ensure that platforms that believe they have invented fire with the advent of blogging and "citizen journalism" recognize that they have merely modernized the pamphleteering of the Revolutionary War era within the graph. A format that exploits mechanisms of partisanship and creates opportunities for individual expression has an important role in society—but is distinct from public information.

Requiring public storytelling to serve as a core function of elected leadership will also render politics and leadership more personal, and will elevate leaders who know the communities they represent. This kind of representation not only embodies our voices but also ensures that our voices are heard, and that the listening essential to healthy public discourse is elevated, along with stories that have been for too long untold. These stories and storytellers are at the heart of reclaiming our diverse voices—and restoring and modernizing our nation's institutions.

# 7

## RESTORE
## OUR INSTITUTIONS

AS WE ORIENT ourselves to the graph and rediscover the power of our own voices, we must not only redesign the platforms we use for storytelling, but also modernize our public and private institutions so they can shepherd and effectively leverage our modern media systems. Companies that manage the infrastructure and build the platforms we rely on to create, distribute, discover, and consume information can step into the responsibility of caring for these systems—and embrace many of the reforms and design principles surfaced in this book. But many of the things we need from those companies require them, at a minimum, to reconsider elements of their business models—and some to reimagine their model altogether.

Almost none of the companies in question are B corporations, and many are publicly traded. If they fail to work toward maximizing profit within the constraints of the market in which they operate, their leadership could easily be sued by shareholders for making nonfiduciary decisions. I still hope the leadership of these companies might see the changes offered here—the ones that build healthier discourse and empower individual agency—as long-term investments in a healthier society that will ultimately be better for their profitability. In the short term (ironically), however, the chronic short-termism of Wall Street probably requires us to take a step further and alter the market constraints under which they operate.

Much as product designers demand and require values-based constraints to make design choices, corporations (from multinationals to start-ups in every industry) need market constraints to formulate business incentives. We must acknowledge that without direction, private markets are not going to provide the public goods we need as a society. Furthermore, we must acknowledge that direction is both cultural (an expression of societal values) and regulatory (a set of constraints intended to guide systemic behavior). Entrepreneurs need the inputs of constraints, too. And regulation will not curb innovation. Regulation directs innovation toward goals that society wants. More thoughtful regulation will unleash our innovators in that common direction rather than allowing them to wander on unintended and undesirable paths.

We have regulated markets now—they are just poorly regulated. Just like our expectations and habits come from our old channel-based systems, markets are stuck with regulations designed for that same old channel-based world. We don't need more regulation; we need different economics supported by modernized regulation. We need to focus on outcomes that will generate more productive innovation.

The media systems that we rely on for the stories that define us and our democracy exist at the intersection of community, innovation business culture, and media and technology regulation. Innovative storytelling has exploded in this very landscape, which is defined by each of these in various parts and to various degrees, and is further defined by our particular moment in history. For the past century, American business culture has been eroding traditional restraints on monopoly power in business, which has in turn expanded corporate power in politics. That combination has led to a culture of monopolistic innovation (companies that seek to own entire industries) and a consistent process of deregulation that accelerated in earnest under Reagan in the 1980s but has continued apace through Republican and Democratic administrations alike.

## EVOLVE THE ECONOMICS OF ATTENTION

As long as there has been mass media—for over half a millennium—we have lived in an attention economy. The volume and velocity of our current systems make their effects on us as individuals and society much more extreme than in the past. The obscurity of how our attention is monetized makes the effects on our behavior even harder to grok and to defend against. Recognizing that stories and storytelling are in fact essential to humanity and to democracy is a starting point. Acknowledging that the companies we rely on for information bear responsibility for how those stories are created, published, discovered, and shared—and the incentives at work in the conversations they create—is the next step.

Much of the transformation we need from these companies begins as culture change within the business community and the companies themselves. New values at work throughout media and technology companies, from leadership to boards to markets, will lead to shifts in business models. If companies embraced these values and were willing to design businesses based on them, we would see whole new models of publishing and distribution emerge.

What if we abandoned the false choice of ads or subscriptions? What if instead we favored new types of transactions (micropayments) or new exchanges (multiparty data transactions), where people participate in the economy of attention rather than being exploited by it? With a micropayment architecture built into open information protocols, concepts like public information credits granted to all citizens, to be used on qualifying civic content, become possible. If we want help making choices, then we should be willing to pay for curation. That is a distinctly different transaction than having choices made for us without our knowledge or getting something ostensibly for free while the real value is extracted behind the curtain of an obscure terms of service agreement that practically no one reads and even fewer understand.

If companies want to use our data to inform those choices, that's a transparent trade we ought to be free (but not required) to make. If companies want to monetize that data in other ways, we should be compensated as part of their supply chain, just like every other input and cost of goods sold. New types of transactions will give people more visibility into value and choices. They'll have more agency over how and in which markets they participate.

Under these types of models, will the Facebooks and Googles of the world make less profit? Maybe. But that fact is a feature, not a bug. The tech giants currently disintermediate and aggregate more profit than the value they create because they have essentially excluded all the other people from the market and they don't pay for inventory. That's not the behavior of innovation and free markets; that's the behavior of racketeering thugs.

Beyond the transaction-level models of building revenue, corporations must continue to examine the long-term value of exploitation as a business model. Increasingly companies are abandoning the shareholder attitude for the stakeholder mentality. But what if they continue to broaden their understanding of their role in their communities and accept responsibility for community consequences they bear? All nodes in the graph could then assume all roles and enjoy the same potential, with nodes of a higher degree (those that are connected to the greater number of other nodes) and greater power (those that control an edge) addressing unique responsibilities that come with that power. Just as we expect individuals to accept the additional responsibility that comes with more power in a flatter information architecture, companies similarly must accept that the attitudes and new responsibilities are necessary in this new world.

We see these new attitudes at work in some companies and unlikely places already. Patagonia has transformed itself into an environmental activist community that sells outdoor gear, regularly choosing advocacy and community (long-term mission) over sales (short-term profit) and always engaging in ways that are available in their community beyond product purchases. Larry Fink, the founder and CEO of Blackrock, has used his past few annual letters to push harder and harder into the base culture of business and finance, where the distance from ground-level effects produces notoriously little awareness among leadership of the unintended consequences of turning one dollar into two. Fink's core recognition goes beyond the responsibility of companies as community participants; he makes the business case that in the long term, only companies that embrace purpose and a broader set of stakeholders will be profitable.

"Purpose is the engine of long-term profitability," Fink stated categorically last year. Expanding this outlook as the standard cultural assumption in corporate America will lay the foundation for better economic models and design decisions in technology and media companies.

## REGULATE FOR HEALTHY INFORMATION

Self-reflection and humility are fundamental to the cultural changes we need to see in journalism and among the platform companies. Yet these qualities alone are insufficient for achieving the task of transforming our media systems. Our markets are not free. They are regulated in ways that shift power toward companies, make monopolies more likely, and skew incentives against the people. Our entire regulatory framework for media rests on the thinking of the previous broadcast era; it all needs to be recast to reflect the rules and consequences of the graph.

The Communications Act of 1934 updated the regulation of radio, organized all wired and wireless communications technologies under a single law, and created the Federal Communications Commission as the singular regulator, "to make available, so far as possible to all the people of the United States a rapid, efficient, Nation-wide, and world-wide wire and radio communication service." In Depression-era, pre–World War II America, widely available public communications infrastructure was seen as essential both economically and for national defense. Bringing the language of "common carrier" to telecommunications also laid the groundwork for seeing these technologies as critical to civic life. Likewise, regulating them to ensure they provided services without discrimination laid the groundwork for the modern conversation about net neutrality.

The explosion of broadcast media in America from post–World War II through the rise of the internet was governed by this New Deal–era law; the rules the FCC put in place to ensure its effective enforcement; and the fairness doctrine, introduced in 1949, which was designed to ensure public access to a wide variety of viewpoints. Broadcast television was interrupted by the rise of cable networks in the 1980s, and the first major amendment to the existing framework came in the form of the Cable Communications Act of 1984 (CCA). With the nation no longer dependent on the scarce public resources of broadcast spectrum, the CCA sought to widely deregulate cable television and pushed the granting of cable licenses down to local municipalities. The result: the explosion of the cable industry and accelerated fragmentation of media.

The CCA required at least one public access channel in every market, but this requirement rarely produced significant public content, as the law did not go so far as to ensure creation and publication of public material or provide sufficient funding. Meanwhile, the fairness doctrine was repealed during the Reagan administration, a move framed in terms of commitment to the First Amendment. That rulemaking focused on the further deregulation of media and, in combination with the expansion of cable, led to the rise of increasingly partisan cable news outlets. Unbound from attention to public discourse or devotion to a full spectrum of opinions, cable news immediately focused almost purely on the expression of commentary and opinion as news.

The cable map quickly became a collection of small, local monopolies that determined the content and pricing for their communities. To protect consumers, the 1984 law was amended in 1992 with a set of pricing regulations meant to reintroduce competition into the system. So began the patchwork updating of an outdated legal framework that greeted the rise of the internet—and has been subsequently turned upside down.

In 1996—fresh on the heels of Tipper Gore's battle over labeling "objectionable" content with parental advisory stickers that primarily singled out (read: censored) Black musicians, starting with the hip hop group 2 Live Crew—Congress passed the Telecommunications Act of 1996, including Title V, which is separately known as the Communications Decency Act (CDA). The broader bill was intended to bring cable and emerging internet regulation into line with the radio and television framework from 1934 and under the authority of the FCC. It largely codified existing broadcast thinking and terminology for the nascent internet while ensuring that the *distribution* of content was separated from the *responsibility* for content. The CDA attempted to regulate internet content for the first time, but (via the now-infamous Section 230) it also ensured that internet service providers were not to be construed as publishers and therefore were not liable for the content of third parties using their systems. In 1998, via the Digital Millennium Copyright Act (DMCA), this liability protection was extended even further for the platforms that would become the core distribution engine of internet content. While ostensibly increasing penalties for copyright infringement online, the DMCA absolved platforms of the responsibility for infringement by acts of their users.

The decoupling of innovation from responsibility has been hailed as the rule that built the internet. It paved the way for unrestricted innovation, including few repercussions for any unintended consequences. But this law did not take the opportunity to express explicitly how greater connectivity might serve public discourse. Lawmakers did not express how the law ought to enrich public discourse, or how failing to envision the proper support of public discourse might ultimately undermine it, or how more access to more information would be good for civic life.

As a society, we assumed the implicit intentions of the internet inventors would come to fruition, without explicit guidance, in the corporate expression of social and ad networks—even as we remained ignorant of how those networks would ultimately capture and monetize our attention. This patchwork of broadcast-era law and rulemaking never truly envisioned the situation in which we now find ourselves: the concept of the graph, the massive decentralization of power, the enabling of asymmetric information. We have a regulatory environment that never defined the actors it hopes to regulate.

Now, rather than creating proper modern definitions and roles for new platforms, we are constantly fitting new systems into ill-fitting historical precedents. In this environment of expanding power and opportunity, no one is responsible for the consequences of that expanding power. No one is accountable for the destruction of our public sphere and civic life, where there is no media equivalent to an "aiding and abetting" charge in criminal law. Instead, we collectively watch and wring our hands while companies that claim to be the most powerful and smartest in the world hold up their hands in capitulation, saying either, "Too hard to solve!" or "Not our job!"

Expanding the liability of platforms as aiders and abettors of social harm is the beginning of a new regulatory framework for media. But removing the shield without clarifying the intentions of responsibility would open platforms to both nuisance and partisan legal challenges that could have the effect of *decreasing* platform efforts to shape healthy community discourse. Because the challenges of healthy information extend well beyond the limits of simple solutions, we must push the FCC and Congress to revamp the CDA—to recognize and embrace broader economic regulation of both media and journalism. This could ensure our media systems provide the kind of information we need, in ways that are accessible and available to everyone.

Updating the Telecommunications Act of 1996 requires a new definition of *common carrier*. It also requires a definition of *media platform* that distinguishes internet access from information access from publishing, in ways that make explicit the roles of Google, Facebook, and the like. An amended act should clearly limit these technology giants' participation in the graph—both how they enable connectivity and how they select what content is consumed by other nodes (us as individuals).

The fairness doctrine was well intentioned in 1949, when it was built around the idea of expressing dissenting opinion on controversial issues. Back then, the media choices comprised three broadcast channels and whatever local print and radio media was available, but there is no way this seventy-plus-year-old law is appropriate in a world of exponentially increasing sources of information. It has been abused to create ideological conflict on issues that are not under debate and to conflate social conflict with scientific inquiry. This blurring of the lines between analysis and fact and between commentary and information has had profound consequences for our ability to make sense of the world in objective terms. It has undermined the role of science as a tool of society that contributes to our understanding of the world.

The suggestion that any public channel should attempt to make space for all opinions has been bastardized to make space for and normalize fringe beliefs and to masquerade those opinions alongside established law and scientific fact. We would be better off focusing on the accumulation of ownership as a limiting force on the choice and power of individuals to experience multiple viewpoints easily. The concept of fairness could be reintroduced into our systems design as a mechanism for ensuring transparency of opinion and bias and could inform a mechanism for identifying scientific fact and for drawing the line between interpretation and synthesis of those facts. Rather than forcing partisan actors to feign unbiased content, we can focus on rules that inform and clarify who is making choices about what we see and hear. We can enshrine rules about the diversity of ownership in multiple dimensions throughout all media markets.

Creation and publication have traditionally been connected to the distribution of content by various media. Newspapers are both the journalists and the physical artifact we read. As creation and distribution are disentangled, disintermediation has increased, and traditional business models have become fragile. Regulation aimed at ownership of publication addresses only a portion of the challenges we face; we also need rules focused on distribution and discovery, which are dominated by advertising (the massive majority driven by Facebook and Google) and by either intent-based discovery (search, which basically equals Google) or serendipity-based discovery (social, which basically means Facebook).

Two companies—neither of which create any content—have come to completely dominate the stories we consume. This speaks to the effectiveness of disintermediation and to the need for rules about leveraging power in ways that serve not only the business, but the public. More active antitrust enforcement is needed across most industries, but it is desperately needed in our media systems. Updating how we think about monopoly relative to information access and diversity of opinion is essential to healthy discourse. Recognizing that our old models of ownership may be insufficient, we may need to develop different regulations for companies that are both node and edge, or that control multiple edges. Seeing dominance play out across the graph will have different consequences and may appear different economically from the abuse of a dominant position in, say, the steel industry a century ago. Focusing only on the price consequences of monopoly does not fully take into account the broader social and civic consequences from lack of innovation and discourse distortion.

The regulation of modern advertising is very limited and generally focused on industry limits based on perceived harm to people. In addition to weak limits imposed by the FCC, the Federal Election Commission puts narrow limits on certain types of communications defined as "express advocacy" by the *Buckley v. Valeo* Supreme Court decision that created the standard, but not the *amount* of spending; by and large, it does not dictate standards of transparency or consequences for misinformation that influences people for political ends. As we begin to understand the dire effects of unhealthy discourse in civic life, these kinds of misrepresentation—both intentional and otherwise—should be bound by *higher* standards, not lower ones.

Facebook in particular has actively avoided any management of political speech, despite marginal boundaries set during the closing stages of the 2020 presidential cycle, by hiding behind the Constitution. The First Amendment was designed to protect us from government infringement on free expression. It has nothing to do with the technology companies accepting their responsibility for the power they express over our attention and choices or helping reinforce healthy standards and guarding against false information that undermines our civic institutions. Expanding liability for content would push these platforms toward that responsibility. Requiring public officials, campaigns, and political advocacy organizations to label content and limiting how they access tools meant to commercialize attention will help narrow the scope of that platform liability, so the story distributors share the responsibility more evenly with the storytellers.

As paid media has evolved, we have banned certain industries from using those tools to reach people. With public health concerns in mind, we have limited their ability to reach vulnerable populations, such as children. Alcohol and tobacco products, for example, face strict limits on where and how they can be advertised. That same logic should be applied to other industries whose access to—and ultimately, abuse of—the models of information and advertising can damage our physical health and our civic health alike. Direct-to-consumer pharmaceutical product advertising is a prime candidate. No less egregious (and perhaps more so) is paid promotion of "advertorial content," which intentionally blurs the line between content types and elicits commercial engagement under the guise of information.

Redesigning interfaces to increase distinguishability and to make intent clear and obvious may make further advertising regulation unnecessary. But bright-line boundaries could act as a bridge to that redesigned reality by helping to ensure distinguishability until removal of certain types of content—those that take advantage of people's assumptions and priming—is accomplished.

We must also ensure that the underlying infrastructure itself—the "plumbing" of the systems—is operated in ways that support the kinds of content and the outcomes we need from media. The general conversation about the fairness or neutrality of this infrastructure is now lumped in with the debate about net neutrality. To the original creators of the internet, the underlying design principles of agnosticism and procrastination were fundamental to the explosion of innovation. These principles have been taken for granted in the second and third generations of connected technologies, and they might have been sufficient to ensure that neither content bias nor class bias existed at the level of data transmission—had they operated outside the realm of commercialization. Net neutrality is not about introducing new constraints to these systems; it simply seeks to codify those constraints that ensured its generativity for innovators who either weren't around or weren't aware of their essential function from the earliest days of interconnectivity.

Last, in addition to its role as regulator, the government must step up and assume its leadership duty to maintain standards. Making explicit the values we expect and demand from media systems is one public step. Defining and issuing open standards that these systems can comply with makes possible the open protocol and pub-sub architecture raised in our discussion about content and reclaiming our voices. As part of the expansion of public content and media, the government should also invest in creating standards of public health for information—defining civic health metrics that our platforms can optimize for, rather than assuming the best-case scenario and then hand-wringing when our needs aren't met.

## REDEFINE NEWS

The graph has changed storytelling. Information does not function as it did in 1920, and the assumptions we carried about authority and credibility that underpinned our ideas about the media and journalism no longer hold. It has been a century since the Columbia Journalism School codified the modern principles of journalism and the ethics of good, trustworthy news. Many of those principles should be refreshed and resurfaced; some need to be rethought altogether; and others need to be included and expressed for the first time.

When these principles were first codified, the idea of individual publishing having the same power and reach as institutional publishing was laughable. Newspaper editors saw the pamphleteers of the nineteenth century as early indicators of ideas and thinking, but their reach was only magnified once larger distribution channels gave them voice. Editors and publishers of the early 2000s saw bloggers in the same way until social media platforms upended their assumptions about reach. The explosion in popularity of podcasting represents a similar challenge and opportunity to reignite the power of radio—especially long-form talk radio—that had until recently required spectrum access and infrastructure well beyond the reach of everyday storytellers. In our new world, where power can increase through organic growth, we all share the same potential for reach, and overcoming institutional barriers has become easier. Combine that with the indistinguishability problem and individual publishing can now carry more weight and achieve more reach than institutional distribution.

The platforms, especially Facebook, are working against this trend because they make less money on it than on content that requires paid promotion. It is in Facebook's economic interest for people with money—not people with the best content—to achieve influence and reach. In a world where even the remarkable must advertise to be discovered, Facebook is incentivized to ensure that stories told by the rich (remarkable or otherwise) are the only ones you hear.

The responsibility of the fourth estate in democracy cannot be overstated: we need journalism to provide the information required for us to be good, effective citizens and for society to function. Nor can the costs of our modern news industry's failure to meet that responsibility be overstated. Thanks to the economics that turned the industry away from its purpose—and due to a failure of creativity in discovering new business models that would align purpose to profits—journalism has abandoned its necessary post. Restoring journalism to this role is absolutely crucial. We need to redefine the definitions and vocabulary of news, restate the principles behind the creation and publication of news, and discover new business models that secure the resources necessary to operate at sufficient scale and quality to fulfill journalism's true purpose.

Journalism has been failing us, but it isn't entirely the fault of journalism. The stories we require in order to be informed citizens cannot compete with content meant for engagement on platforms optimized for outrage and confirmation bias. That cute cat video is always going to outperform that long-form investigative report. Newspapers never used to have the same economics as entertainment. The moment in the early nineteenth century when newspapers abandoned their core purpose—to inform and persuade citizens—and began to strive for mass audiences to reap greater profits, they began a slow process of abandoning their responsibility as our fourth estate and embracing the business model of entertainment instead.

That shift was also the moment our news parted ways with political party patronage and became bipartisan or nonpartisan. This divorce may have been a positive development, but it was simply a consequence of commercialization—not a primary intent. Restoring the focus on fact-based rather than opinion-based narratives and optimizing for stories we need rather than headlines that drive clicks will enable journalism to reclaim its necessary place. A fourth estate that lives up to its name will help elevate our democracy by ensuring we have the potential to be better-informed citizens. These goals may not be as profitable as stories that titillate and entertain, but they are more necessary.

A restatement of intent and principles will help us consume information more effectively and support the institutions that create the content we need. Rather than just an attempt to reclaim old ground, the restatement must stake out the principles necessary for journalism to serve society effectively now, in today's world.

A focus on unbiased journalism, for example, would be fruitless. Journalism today should be focused on transparency of bias; commitment to facts; establishment and promotion of credibility; recognition of authority; clear provenance of data and sources; and intentional distinguishability of information, analysis, commentary, editorial opinion, and external opinion. Furthermore, in a world where citizens on social media are often the first line of reporting for major events, the news media should prioritize context and accuracy over speed. The key here is to make explicit the principles that the graph has undermined or that we have taken for granted in a world where gatekeepers no longer govern our information.

Gatekeeping—which leveraged curation as a tool of civility and, ultimately, suppression of dissent and oppression of minority power—is not the role we want journalism to restore. In our new model, curation can and must be a tool of liberation and empowerment, of expanding viewpoints and sharing power. Just as democracy is a system of borrowed power—individual citizens lend our power to elected officials, who wield it on our behalf—so must editors see themselves as borrowing our power of exploration and helping us discover new perspectives, new voices, and new worlds.

In addition to the expression of updated principles and the shifting of operations and cultural standards inside newsrooms and publishers, the FCC and journalism trade groups could collaborate to create metrics for authority and credibility that travel with each piece of content, each author, and each publisher. This system would allow authority and credibility to aggregate and build public memory over time, as we develop incentives and consequences for our language and for the content we create and share. Accidentally sharing bad information should not permanently destroy someone's credibility, but a habit of sharing misleading or misidentified content ought to be accompanied by a clear and obvious message to people trying to make sense of the content coming from that source.

A right to free expression does not give us a right to consequence-free expression. A lack of public memory undermines our comprehension of the information we consume and prevents us from making good decisions about what we want from our storytellers. Maintaining a ledger of authority and credibility over time could be a positive use case for a blockchain and could be securely and safely made public. This kind of publicly transparent metadata about the information we consume is how we begin to get enough information about our information so we can safely extend our decision making beyond the filter bubbles that feel clear and safe. We never had this much metadata before because credibility was assumed based on the publisher's reputation. Parsing a more complex set of contexts is part of the responsibility we must accept in a world without gatekeepers. In that way, we can restore our agency, making our own choices about what we consume and what we trust rather than abdicating to gatekeepers, platform algorithms, or our neighbors in the graph.

## TREAT STORYTELLING AS LEADERSHIP

As I mentioned in my earlier plea to invest in public stories, we must expect our leaders to embrace storytelling as leadership. We must also demand that the institutions they lead and the structures that give them power make commensurate shifts.

Conflict-obsessed news and outrage-optimized platforms make campaigns and elections the most visible aspects of public leadership. In the Citizens United era of campaign finance law, political speech is exploding and coming from more and more varied sources. The Federal Election Commission has permitted an environment with no accountability or consequences, and it must be drastically reformed to ensure that the standards we need for public stories can be enforced. Reducing the proportion of public storytelling that comes from campaigns would also help us to rebalance our civic life—away from the contest for power and toward the everyday work of self-government.

Rulemaking in Congress is an expression of cultural values. Building new requirements for public dialogue into congressional responsibilities will help us standardize our expectations rather than deferring to outliers of greatness and setting the standard behavior essentially at zero.

Representative Ocasio-Cortez has been recognized on all sides as a transformative leader when it comes to the consistency and openness of her storytelling. From the beginning, with her unlikely campaign to unseat a white, male, twelve-term incumbent, her ability to connect directly with citizens and to share an honest, unvarnished view of herself and her worldview has shifted expectations for what leaders can sound like. Her eagerness to open up the governing process has created and renewed interest in the essential but often unseen processes of governing and public leadership. But we need her habits to be the norm.

New standards and templates will be required in order to pull Representative Ocasio-Cortez's less capable, more reticent peers into the twenty-first century. They are going to need encouragement to help them understand more clearly what healthy public leadership looks like. As for those who prefer to operate in the comforting, corrupt, murky darkness—they will have to be forced into the light or become obsolete.

## TRANSFORM ORGANIZATIONS

While elected officials are significant and important sources of public information, they are far from the only ones. Transforming public institutions to require and refresh storytelling as a fundamental pillar of leadership is not enough. This step must be mirrored in our private organizations, by making media a fundamental pillar of public literacy.

Public institutions anchor our civic stories, but private companies and nonprofit organizations, which interact in both public and private markets together with regulators, have their own stories to tell and their own role to play in ensuring clear, complete public information. Embracing their role as storytellers—as essential sources of information about themselves for the people they serve, employ, and transact with—will require, first, a shift in mentality and culture, and second, a shift in organizational design.

The first shift is the recognition that the change in architecture of information—from channels to the graph—also shifts the organization's posture of storytelling from broadcaster to participant. Unfortunately for companies and organizations, which are used to the comfort of being the center of their own world, this shift in perspective can be profoundly jarring. Yet it amounts to a recognition of reality more than an innovation. Recognizing the ongoing conversation about an organization's brands and work extant beyond its four walls (yes, without its permission) is an opportunity to build more relationships with people who are already engaged in its brand narrative. This reality may be less predictable than the company's old models, but it represents more opportunity than the company has ever had.

This opportunity requires that organizations treat these relationships as relationships within a community, not as an audience to be targeted. Listening—to be in conversation *with* people rather than talking *at* them—is the most important fundamental skill organizations have never had to learn. The graph presents this opportunity, but also demands something in return: companies must recognize that not only do they wield less control over the conversation about themselves and their brand, but their community members take on greater power to drive that conversation and expand the community. The good news is that consumers generally are a more trusted source of information about organizations than the organizations themselves are. The bad news (again) is a lack of control. The second shift in transforming the organization to take on a more participatory role demands that it see people as people, lead its communities with values geared toward shared benefit, and make the community stories as rich and purposeful as the ones the organization tells about itself. This shift forces major redesigns of staffing, capabilities, and organization to enable very different behaviors and habits. With these redesigns, all communication becomes an exercise in community organizing—recruiting, enabling, and building power among others in service of common goals.

These brand narratives are often about selling products or services, but increasingly must include more attention to the community and its concerns. A broader recognition of our interconnectedness will go a long way toward credibility. The reality is companies cannot segregate their marketing identity from their operational values and then expect people to pay no attention to the man behind the curtain.

Publicly traded companies and other large private brands have been working toward new and more transparent metrics that go beyond financial data to encompass everything from environmental impact to governance to sustainability via new standards groups like SASB and B Lab. These movements have helped to expand organizations' understanding of the role of business in community today, and one can track the slow evolution in how more traditional corporate trade associations view success. Rather than simply working from a system of punishment and avoidance, these standardizations help organizations make explicit the values they expect from corporate and organizational actors in society. With a healthier cultural posture in place, the organization then needs to build the capabilities and capacities to embrace its new values. For nearly all companies, this requires an evaluation and redesign of the organization.

Making listening and storytelling native to an organization means occupying its space in a public conversation. It demands the development of new capabilities and capacities unlike those we've ever expected from the organizations we build.

All of these organizations embracing their role as storytellers also creates an opportunity for us as individuals to hear more, share more, and take a more active role in dialogue with the institutions we engage with day to day. A civic life enriched by increasingly diverse productive discourse, informed by news designed to inform, and with a candid view into active public leadership is on the other side of this restoration. These opportunities also require us as individuals to embrace our powers to listen, to hear, and to share stories—and the responsibilities that come with those powers—in this amazing system of information.

# 8

## REDEEM OURSELVES

F MORE FREEDOM and more information are to be the fundamental building blocks of an architecture of storytelling that offers us more power and more opportunity, then we must embrace this new landscape completely, without holding on to old structures that no longer serve us. We must be bold in our convictions. How do we want these systems to serve us? How can we guide them away from exploiting us? How can we lead confidently in a new direction?

For the American people, stepping back into our civic life is an act of forgiveness: we forgive our leaders who have abused and corrupted the systems. We forgive each other for falling victim to the easy, cruel tyranny of outrage. We forgive ourselves for ceding the arena to misguided players and nefarious actors. But we—the people—are the most common type of node in the graph. If we change, the system changes. Period. And that change must be guided, intentional, and grounded in a concrete understanding of the architecture within which we tell, share, discover, and hear stories.

The exploding pace of innovation is dizzying. When Yahoo first launched in 1994, there were fewer than three thousand websites. Twenty years ago, there were just under 18 million. Now, there are more than 1.6 billion.

Fifteen years ago, Facebook was still headquartered in a Harvard dorm room. Twelve years ago, there was no iPhone, no Instagram, and no Spotify. Five years ago, TikTok did not exist.

Our experience of this exponential evolution has largely been one of passive wonder: looking out in awe at a glittering new world of technologies, accepting "free" experiences that feel expensive emotionally and spiritually, feeling exhausted by the pace of innovation. But the connection between that exhaustion and the "free" experiences is difficult to make, and we are left confused, waiting for our cues to re-engage and for new rules to make that engagement healthy.

The truths about information that we expected to be self-evident were not. Our passivity has left a vacuum of values and direction, in technology and in politics, that has been filled by commercial and narrow self-interests. When we cede technological evolution to corporate, return-seeking capital, of course the platforms will be exploitative and optimized for profit. When we cede the political swamp to swamp monsters, of course politics will feel terrifying and corrupt. It is the absence of light that makes these spaces dark, not that they are inherently shrouded. And if our public spaces are not inherently broken—not inherently incompatible with the kind of open, pluralistic, vibrant public society we crave—then they can be fixed. But we cannot count on the swamp monsters to fix them. We need a regulatory renaissance, a restoration of our institutions. Our public sphere can only be fixed by us.

## MORE FREEDOM + MORE POWER = MORE RESPONSIBILITY

It turns out Spider-Man's Uncle Ben was right: with great power *does* come great responsibility.

In our new architecture, roles have become unfixed, and each individual's potential is dramatically expanded. The guiding principles and pillars of understanding are untethered from gatekeepers who once controlled the subject and flow of information, leaving us unprotected from fringe content. In this architecture—the graph—we have to take on the responsibility for our actions and choices. We have to provide the discernment of authority and credibility for ourselves.

We have ceded our decision making to algorithms in our current platforms that are not optimized in our interest, that have substituted proximity and confirmation bias for the pillars that make true understanding and discernment possible. The former gatekeepers generally abused their power within the system to reinforce white male hegemony and to push anything counter to dominant culture, not just the unauthoritative and incredible, into fringe categories. Curation was a weapon of oppression wielded under the guise of editorial judgment and quality. We must take control of the tools of curation so they serve us, both as individuals and as a society.

Too often, conversations about personal responsibility in politics and in media end up being finger-pointing sessions. We punish people for dysfunctional politics, heap blame for bad leaders onto "apathetic voters," and lament the dearth of quality and healthy media. We scapegoat the viewers and parrot the publishers, who claim they are "simply providing what people want." But "personal responsibility" is a dog whistle for letting exploiters off the hook, an excuse to serve society's lowest instincts. In an architecture where people have more power than ever and more potential to express multiple roles (from creator to publisher to distributor) in information systems, we must accept that with the increase in power comes an increase in responsibility for all the nodes.

That responsibility is the foundation of our new power.

The responsibility to be our own gatekeepers—it seems simple enough. But it actually requires that we build new skills in skepticism, new literacy in media, and new muscle memory so we can tell the difference between being uncomfortable (read: challenged) and being threatened (read: unsafe). The difference is deeply personal. What feels challenging to me as a straight, white male may be emotionally, physically, or spiritually unsafe to someone else. But in all cases, the distinction matters for our ability to remain in community with others with whom we disagree.

The collapse of gatekeepers has given rise to the explosion of voices while taking away the gatekeeper's function as arbiter of authority and credibility. Ultimately, the increased diversity of voice and the expanded access to more storytelling for more communities are wildly preferable to the limited perspective of the past. As authority and credibility are necessary features of functioning media and information systems, however, we must find ways to implement them that are inclusive by default.

Both platforms and institutions can play a role in identifying and distinguishing credible content sourced from authoritative storytellers. In the graph, ultimately, each node will still have to decide for itself: What is credible? Who is an authority? With better signaling and with less substitution of proximity and confirmation bias, the graph will indicate less divergence (despite our modern state of uniqueness) about what is credible and what is not credible. And increased signaling also produces better shared definitions of credibility and shared badges of authority that can travel across platforms and unite broader swaths of society into trusted conversations with shared baselines.

Still, with billions of decision makers, we will have billions of decisions. In order to create a shared space for healthy discourse, we also need to accept standards and rely on open validation and accreditation mechanisms. This will ensure that we have at least some overlap in that decision space and that we can leverage these systems without succumbing to exhaustion.

Making good decisions about information also requires greater understanding of the systems we rely on for information and willingness to assess and evaluate content with thoughtful skepticism. It is essential that we do not confuse reflexive cynicism about others and about challenging information with that healthy, productive skepticism we need.

While we can look back and see storytelling as a fundamental human experience, the storytelling tools we depend on in today's world and the complexity of the systems our society uses for the creation, publishing, distribution, and discovery of stories are based on more than an instinctive understanding. The complexity and power of our current systems—and our asymmetric grasp of them—are what give bad actors so much power in our current landscape.

Propaganda, misinformation, and disinformation are nothing new, but today they are more effective than ever. Thanks to the speed of our current systems and the expanding delta between power and understanding, we are living through an era of maximum manipulability. Building our public media literacy, beginning with treating it as a fundamental pillar of literacy for children in school curriculums (see Megan Kiefer's work at Take Two Film Academy), is necessary to rebalance this imbalance of power and reclaim our agency as individuals in the graph.

Then, as educated, thoughtful content consumers, we need to get curious about the world around us. With the safety and security of greater understanding, and with a better sense of brave versus safe spaces (hat tip to The People's Supper and Micky ScottBey Jones), we can begin to reach out from our filter bubbles, intentionally seek out stories that expand our perspective, and interact with a world of greater connectivity. In a new environment that makes little sense, where our basic markers for society are unrecognizable, it is a pretty natural human reaction to turn inward toward confirmation bias for safety and predictability. But only when the systems themselves increase our understanding, allow us to distinguish the useful from the distracting and manipulative, and give us a greater capacity for productive discourse can we begin to embrace the potential of greater connectivity and humanity.

Right now, that experimentation may feel like wandering into a dangerous jungle. The least vulnerable among us must remember their responsibility for confronting the bullying and abuse that shift *challenging* to *unsafe* for too many people. Yet our media platforms hold the promise of support rather than exploitation. With better markers and with systems and norms meant to help rather than exploit us, our exploration can become thrilling and generative. We can hope for a future when we can rely on healthier systems designed for the purpose of healthy discourse.

In the words of Uncle Walt again (or Coach Lasso, if you prefer): we can be curious, not judgmental. We can strike out for new ground, beyond the confining safety of judgment and the false sense of security that comes with cynicism. We can bravely open ourselves up to *not* know, to be uncertain, to ask questions, to discover.

## EMBRACE SHARED EDGES AND THE NEW PUBLIC SPHERE

Central to our role as explorers is our willingness to embrace shared spaces—to intentionally seek out environments where we will actively interact with people unlike us, on any number of dimensions. As the Civic Signals project has discovered, sharing physical public spaces leads to dramatic increases in civic participation and to reduction in crime, largely as a function of building familiarity (not necessarily relationships) with more and more different people. We need digital spaces and storytelling spaces where diversity and familiarity with that diversity is the norm, where the default experience is inclusion. We need to build habits of spending time with others, exploring the market, and walking down Main Street together.

In 2018, I left downtown Chicago after almost twenty years in that city and moved to a small village of fewer than 2,500 people in the Hudson Valley region of New York. My wife and I moved for family, but also to reclaim something we were missing but couldn't clearly identify. Despite a huge and loving network of friends and colleagues, a pickup soccer game I knew I wouldn't be able to replace, and twenty years of history in Chicago, there was something increasingly isolating about our urban life.

When we got here, we were embraced. I joined the volunteer fire department and became an EMT just as the first wave of the coronavirus pandemic was sweeping through the country. My wife was a regular at town meetings, joined the zoning board of appeals, and quickly became a village trustee. We go to the farmers market every Sunday. We became locals (as opposed to weekenders or transplants) very, very quickly because we are active and visible, because we sought out community. We are part of the public fabric of the town. And importantly, there is a public fabric of the town to be part of.

In much of modern discussion about society and social relationships, there is a hard line drawn between online and offline interaction. As online experiences grow, many lament the perceived collapse of relationship quality that results from too much screen time. We worry about our failure to connect with people face-to-face. But while there is valid and important research into the differences (neurological, sociological, and psychological) of screen-based experiences versus in-person experiences, the instinct to reject online automatically as the lesser of the two is a nostalgic and misguided interpretation of humanity and society. Reflexively embracing all innovation as progress is equally misguided.

The key is finding a new sense of balance and clarity. How do we value new interactions, and which types do we value? Which experiences weave us into a public fabric, even in a pandemic era that demands physical distancing? What do those experiences make possible for us? What do they limit? This new balance is a complex and necessary calculus. And listening to scientific research on the effects of new systems is important if we are to retain our sense of agency and proactive choice—especially when we're up against a multibillion-dollar industry that would prefer to make choices for us about what we attend to and how we interact.

If we want systems to serve us, we must demand and use individual control over experiences where systems of storytelling are optimized for users like us (i.e., individuals), not only customers (i.e., advertisers). The regulatory transparency and guides described in the previous chapter will help make that power more easily accessible to us. Still, it remains up to us to claim it, use it, and remain vigilant in our embrace of technologies that serve us and society, rather than acquiesce to technology in service of itself or of ends that are outright antisocial.

Despite our sense that they are "free," none of the platforms that we have become accustomed to actually cost us nothing. In addition to the hidden direct costs, the second-order social costs are profound. Rising extremism, decreasing social cohesion, and corporate control of our cultural narrative all undermine us as individuals, make our relationships more difficult and less healthy, and reduce our capacity for effective self-government. These costs are spread across society, deferred by the platforms, and amortized over time, so they are hard to attach to their original source. But make no mistake: these effects are the cost of exploitative information systems and our continued use of them.

If the stories we share and believe about ourselves form the foundations of society, then creating, controlling, and changing those stories is power. What we believe, what we attend to, and what we share are expressions of citizen power in a democracy. Our casualness with that power makes us complicit alongside the platforms and regulators that are either ignorant of the damage they are inflicting or outright malevolent and corrupt. At a minimum, we cannot assume that our implicit needs and desires for society will be met by these systems. More practically, we cannot afford to cede our power to the corporations or to the platforms themselves. We cannot allow our shared stories to be dictated by their business models. We have more agency than we claim. Empowered with real choice, we are capable of this transformation.

## BUILD NEW HABITS—ACT AS IF

Storytelling is an expression of power. So how we tell our stories, where we consume them, and which ones we consume are not only expressions of our power, but also tacit acceptance of the storyteller's power to influence us. The choices we make about each of those things are our expressions of power over how we want the world to work for us. But it is easy to fall prey to these systems, to fall into click-holes of near-catatonic consumption; it is easier to blame the corporations, platforms, and failed leaders for exploitation. The fact is, we have been losing the fight against the multibillion-dollar media industry, but with the help of redesigns and restored institutions, and with our concerted effort, we can change our trajectory.

These systems are intended to be habit-forming. We are, in fact, addicted to them. They are designed with feedback loops meant to prey on our brain chemistry for rewards and with outrage to keep our attention. They are configured to keep us clicking on more and more content, regardless of whether we are informed or satisfied. Being informed or satisfied is not the goal of these systems—it's not how they are optimized. In fact, they are built to ensure that we *never* feel a sense of completion or "enoughness," so that we keep searching, clicking, watching, playing the game, and driving the engine of their attention inventory.

As with any habit, breaking our information routine is hard. Humans are used to a steady diet of hyperpartisan outrage, and suffering from dopamine withdrawal is a recipe for continued addiction. Beginning to see the costs of these habits and starting to connect our personal costs to the broader sociopolitical ones is the rational foundation for beginning to adjust our behavior. Emotionally, we must also recognize what we are getting from these bad habits—and what we are losing. Confirmation bias is affirming. And in a lonely, isolated modern life, affirmation is a powerful drug. This truth is made worse by the recent pandemic realities, especially our lack of access to others and our inability to gather. Blame in the form of outrage, as an explanation for shortcomings and frustration, is just as appealing. These feelings are natural reactions to being exploited, to our frustration that leadership cannot seem to help us change direction. To begin shifting our behavior, we have to be willing to risk the discomfort of letting go, of releasing the balm that outrage provides. Our comforting filter bubble is only a stand-in, not a sense of genuine belonging. It robs us of real community.

In order to begin, we must first allow the future we want for ourselves—not our seemingly broken, frustrating past—to drive our present.

Information we need to make good, complex decisions is readily available right now from diverse, credible sources. Entertainment that excites and inspires us and that also opens our perspectives to other mythologies than the ones we may have been raised in make the world feel smaller; hearing and seeing stories told *by* people themselves rather than *about* them makes a world of difference. Leaders we know, who listen to us and their communities, who engage daily and give us clear windows into their work and their thinking, who invite us to be part of building communities, are there to engage; they call us to support a country that reflects our values and has public systems that do what they can to drive thriving communities for everyone.

This future may feel like an impossible product of our present, but it *is* possible in a world where storytelling is for all of us. The stories that we collect to define society and ourselves can either unleash our creativity and wonder or bind our ambition into a cruel, narrow futurelessness. Ultimately, it will be up to us to choose the stories that free us and guide us toward the future we crave.

## LEAD

We were promised jetpacks but given cat videos. The companies we rely on for information discovered, without cultural direction, that cat videos were more profitable. With new direction, it will be amazing to see what those companies provide, what that dizzying innovation engine will create in service of principles that support rather than exploit society.

Our platforms need redesign and our institutions need restoration, but that work is going to be done by people taking the reins in those organizations and changing first their culture, then their behaviors. Many of the leaders of these organizations want them to be healthier but need a road map to get there. Some others will hold on, kicking and screaming, to their exploitation engines of profit. In either case, together we can provide the leadership necessary to begin a new era of evolution in these information systems.

No one else is coming to fix these things for us: *we are the cavalry*. So step one is showing up for the conversation, educating ourselves on what's not working and why, and beginning to envision things differently. The expansiveness of the systems we rely on for information and storytelling is daunting; pondering where to begin can make us feel paralyzed. But the scope of the change required can be freeing when we realize that we can start anywhere. There is no perfect order of dominoes required to begin this new evolution.

So we begin where we can, wherever we are inspired, wherever there might be leverage or opportunity to begin the shift. Try new platforms. Create new platforms. Use the time management tools built into your phone and computer. Keep your phone on Do Not Disturb *all the time*. Turn off all your notifications, with a few carefully selected exceptions. Unplug from the dopamine traps. Support new tools with better ethics. Back advertising boycotts aimed at demanding that platforms embrace more responsibility. Vote for leaders who understand and take these questions seriously. Work in government. Run for office.

## UP TO NOW . . .

When it comes to judgment over curiosity, we tend to get what we expect from others. The further people are from us in the graph, the more likely we are to interact with them as stereotypes or caricatures rather than individuals. We need to choose curiosity. We need to stretch, to push our idea of *family* out beyond our narrowing definitions to include all we can reach.

Our reach has expanded in the graph, but our trust in humanity has not. Up to now, we have greeted more access with the same passive wonder with which we greeted the dizzying pace of innovation, leaving us uncertain or unaware of how to claim our power. We've been waiting for our cue: How can we best take advantage of this new world? How can we understand which parts of it were meant for us? How can these promises be made real? If we want to redeem our civic life, what comes next is active, and it requires us to let go of the protective posture we've held up to now.

Building familiarity (if not intimacy) with more people and more diverse voices will allow us to build more confidence, comfort, and resilience. Exploring and engaging with more of the world will reduce our fear and uncertainty. When I moved to Rhinebeck, I could have avoided the firehouse because the members are mostly old-school Hudson Valley Republicans who might not love my liberal Chicago politics. But I joined because I wanted to be useful, and I found that we are connected by our willingness to wake up at 3 a.m. to help our neighbors. And I discovered that every single member of the department has something to teach me, something important to contribute to my life. As a result, my understanding of the world is immeasurably more interesting, more complex, and more complete.

As we look out on the world and interact with more and more voices, we should be hard on ourselves but easy on others. We should maintain a critical lens on sources and credibility but be generous with credit and intention. There are bad actors, but they are the exception. Treat people as if their intentions are as good as yours, even when they make mistakes. If we expect good from people—and if we want those who are making mistakes or with whom we disagree to find new paths—then we need to make space for them to change. To shift from one space to another, first it has to be possible. Listen for the signal in the noise, even if it is buried.

The balance between healthy, critical skepticism and lack of trust is complicated by these systems and, in particular, by interaction at a distance. When thinking about the distinctions between online and offline interaction, the challenge and limits of experiences without visual, nonverbal cues and face-to-face communication are a major component. Healthy discourse requires that we find ways to argue in safety, to allow for disagreement without making space for oppression. The danger of opening the door to dangerous or unwanted thinking and of being willing and constant in our confrontation of bias and oppression means that we need to focus not on being civil, but on being kind. Civility is too often a tool of the status quo, of silencing dissent, and of exhorting challengers to "be polite." Civil often means compliant. Kind, on the other hand, means decent . . . respectful . . . human.

We should go easy on each other, but our leaders need us to hold them accountable to the responsibilities of their positions. Public leadership must be public, and that means open, accessible, and visible. It also means telling stories about the work, about the process—making service and governing vibrant and relevant by highlighting its importance and revealing how complex decisions and policies touch our lives. A public policy degree should not be required to engage in self-government. It is the responsibility of our leaders to occupy better spaces in restored institutions and to become better storytellers.

We know what the past decade of civic life has looked and felt like. We don't know what's next. The unknowable future we paint for ourselves is exactly that—unknowable. What we *do* know is that a future guided by public principles and grounded in a commitment to public discourse and vibrant civic life will create an environment of innovation and exploration where we will invent and discover things that serve those principles. New things *will* emerge in the spaces created by removing the things that don't serve us—we need to trust in that certainty. That trust is rooted in being explicit about what we want and in knowing that clarity will enable us to create systems that are not simply chain reactions of human emotion, where collision creates heat and reaction with no goal or limiting reagent.

The gravity and weight of our dysfunctional present demand that we actively work to inspect and improve our systems and to shift our civic life toward habits, norms, and expectations that serve us. If we fail to do so, then we are working against ourselves.

The status quo is not neutral or static; its momentum pushes us toward the further fracturing of society and the dissolution of the republic. But culture change is tricky: it seems impossible, shifting glacially, almost imperceptibly, inch by inch over generations, then suddenly . . . all at once.

The decline of our civic life can be reversed by the emergence of this new future that we crave for ourselves and for our country, if we forgive and begin.

# MANUFACTURING
# OUTRAGE

THE MORNING OF November 4, 2020, was nothing like 2008 or 2016. Nothing was resolved—no awakening, no catharsis, no shock. The pandemic year that gave us an April that lasted six months was now treating us to an Election Day that would last a week.

In many ways, 2020 extended the damaged, disconnected, and divergent civic experience of 2016, but because the 2020 election took place during the COVID-19 pandemic, it also presented new challenges to both participation and information. We experienced this latest presidential cycle while also sifting through misinformation and disinformation about voting rules, quarantine regulations, Zoom school, and a multitude of other stressors. And like most stories, the conversations we had with one another and ourselves in 2020 fell into familiar narrative patterns that are hard—but not impossible—to change.

The concept of narrative change work is not new. It is rooted in the idea that the stories we hold in our minds about society and social issues are the foundation for whatever solutions and policies we'll support to address social challenges. With the rise of cloud computing, big data, and modern machine learning text analysis has come the added possibility that narrative change can be modeled, quantified, measured, and tracked. The Narrative Observatory project at Harmony Labs, where I work as community director, was created to build exactly this sort of data-driven narrative infrastructure. It tracks the growth of social narratives over time and provides leaders with the tools they need to impact (and to understand the impact of) critical conversations.

As part of the Narrative Observatory's work, we observed an explosion of news consumption in America in 2020. And while the divergent media consumption habits of liberals and conservatives remain at the foundation of each community's information diet, nearly all Americans found themselves, either directly or indirectly through the graph, consuming some of the narrative presented by edges like the *New York Times* and CNN as we searched for some semblance of coherence about public health. Yet by the time the election was called for President-elect Joe Biden on the morning of Saturday, November 7, opposing narratives about the election outcome had already cemented in the minds of these distinct communities. We were accidentally getting much of the same content, but interpreting and reacting to it in completely different ways.

The common social norms and frames provided by shared experiences and stories define the structure for not just how we receive information, but also how we interpret it. So in today's reality, not only has our experience of the graph diverged, but our frames for interpreting information are so dissimilar that even accidental convergence doesn't result in shared conclusions. The collapse of authority, authorship, and context, especially in the news, means that our reactions and those of our closest communities—and not the content itself—anchor our responses.

We have become so well trained, so habituated to outrage and hot takes, that we have ingrained these automatic, collective responses. We are no longer thinking for ourselves; instead we automatically fit new information into reflexive, self-reinforcing narratives. The new habits of the graph and the narratives of political conflict sanctioned by leaders who are focused on power over service have trained us so successfully that regardless of the intent of journalists or other storytellers, outrage is now our standard response. All of us are stuck in a dangerous groupthink that makes it even harder to relate to people who think differently from us and whose lives don't look like ours. Much of the post-election political analysis and the language of President-elect Biden himself has centered on healing our divided nation, but it is clear that substantial effort will be required across all our media systems to bring America back into a functional public sphere in order to make that healing possible.

In his most famous missive, *Manufacturing Consent*—published in 1988, well before the rise of the internet—Noam Chomsky ultimately concludes that media systems are tools of inequality, designed to ensure state control over a compliant public that consents to their own exploitation. In short, the media maintains the status quo in defense of the interests of a dominant elite. What we have now, however, are media systems that cultivate the opposite of compliance—that are at best disinterested in and at worst actively working against any kind of collective consent, instead favoring conflict and outrage. Rather than make us amenable to exploitation, our media systems keep us in intractable conflicts that prevent us from engaging actively and productively in civic life at all—or turn us away from politics altogether as we cede control of civic life to political elites disconnected from our needs.

The day after the 2020 presidential election was called for President-elect Biden, the *New York Times* ran a wide-ranging interview with Representative Alexandria Ocasio-Cortez, who promoted the uncontroversial idea that it was time for the Democratic Party to invest in its own resilience and innovation. Bereft of the Trump Conflict and Controversy Machine, reporters from nearly every mainstream outlet from the *Washington Post* to MSNBC to Fox News reported this interview as a rift, a schism, a war between antagonistic wings of the Democratic Party. We cannot dismiss this hyperbolization of productive discourse: it is the very problem we face. And it is not just a function of social media.

In his acceptance speech, President-elect Biden spoke of the need to turn down the temperature in our arguments over politics and current events. This imperative applies to professional media and journalism just as much as it applies to political leaders and to us as individuals. Our parties and our political campaigns must put away the language of violence and war if we are to have any hope of engaging as opponents, not enemies. Yes, emergent technology platforms have rearchitected our information landscape into a graph, but nearly all the edges in the system now share the incentives those platforms have introduced. Media outlets and journalists covering politics and civic life must avoid the cheap, easy focus on conflict. By the same token, we—their audience—must give up the righteous indignation and existential posture toward disagreement.

The *New York Times* has never had more subscribers and seems to think that its business model problem has largely been solved by the Trump presidency. But without the Trump conflict machine, journalism will need a new outrage engine to drive its economics. If another Trump-type figure does not emerge to provide it, journalists will end up manufacturing the outrage necessary to stay in business.

There are clear examples of journalists stepping away from the outrage machine, such as when broadcast channels like NBC, ABC, and CBS refused to air President Trump's speech on Election Night. In the days after the 2020 presidential election was called for President-elect Biden, even Fox News began to cut away from White House press conferences, openly stating that the network could not "in good countenance" air false statements or unsubstantiated claims. But CNN anchor Jake Tapper (who usually is finely attuned to the media lens of accountability and metadialogue) simply referenced the steps taken by Twitter and Facebook to label as misinformation President Trump's claims and rants, before proceeding—while covered in real-time chyrons—to claim his network had no capacity to do anything similar. With public consumption of news still dominated by television, TV journalists who point to social media as *the* problem are further abdicating their responsibility. Traditional media platforms bear as much responsibility as social media platforms for what they enable. The unusual decision to cut away from a sitting US president proves that networks do have power to stop the spread of misinformation and disinformation—if they are willing to put healthy civic discourse ahead of profit.

Covering disinformation and misinformation is exceptionally difficult. It's irresponsible for journalists not to cover what the president of the

United States is doing and saying, yet equally irresponsible to extend the reach of falsehood, whether unintentional (misinformation) or intentional (disinformation). The key to this effort, as pointed out above, is the phrase "in good countenance," especially in the context of the profits made possible by the tyranny of outrage. To ensure media outlets make these values-based choices—moral choices of conscience—we the people must work harder to express our values with clarity and conviction, so journalists can return to the work they crave and the function we require from them. And these values can give media companies real incentives to respond to our civic needs over their business goals.

Misinformation and disinformation have always been features of our information landscape, but the graph we inhabit now has been reinforced by the processes that create falsehoods and make them harder to distinguish from truth. Misinformation and disinformation are societal and storytelling problems that speak about trust and sources, about shared experience and filter bubbles, about confirmation bias and lack of access, about the way conflict and outrage drive the monetization of our entire information life. Every node, every edge, every story, every institution within the graph that transmits content, sets our rules and norms, and consumes information must accept some level of responsibility for our dysfunctional public sphere.

Professional media systems must learn to *cover* misinformation and disinformation without *spreading* it. Public institutions must learn to tell stories with the same graph literacy and conviction as entertainers. People must learn to consume content from each edge with the healthy skepticism necessary to understand the world accurately. We must engage with false information and recognize it as a social problem desperate for a solution. Only then can our society continue as a viable democracy, much less a vibrant one.

The 2020 presidential election cycle dramatically (and painfully) reinforced two key realities: one, we need to design media systems that manufacture healthy civic discourse; two, once people are honestly, effectively informed, we must trust them to make educated, complex choices and to govern themselves. In order to accomplish either objective, we must first understand more clearly how media systems work for and against our civic needs. Then we must embrace the conversation about how these media systems work and for whom. If we value a functional civic life over outrage, and if we hope to share an America that can move in a direction of inclusive success by elevating public governance, encouraging productive conflict, and empowering constructive conversations about our collective future, we must start this work today.

Rules and norms are beginning to shift. New momentum is taking shape as, for example, the tracking cookies that enable the entire cabal of digital paid media are being limited or eliminated, and political ad targeting is being curtailed. These shifts, like the solutions I have proposed in this book, make abundantly clear that our future can be one less dominated by paid media, and one where authentic, organic storytelling is ascendant.

Part of these organic conversations must be our common expression of values—a moral clarity that will shape our future. Values-based expectations, including explicit communication of the public goods we demand from all our systems, are necessary not only to address our media platforms but also to serve as a framework for even more significant advancements in technology. We sometimes belittle these arguments as simply a function and a by-product of social media, but no one can deny the implications of artificial intelligence, synthetic biology, or autonomous weapons. In the same way that we must confront the runaway commercialization that has made efficiency and growth the inherent goods of modern capitalism, we cannot allow a hollow, technocratic approach to public policy and regulation to constrain our humanity. New modern media systems can serve our civic life rather than be designed to exploit our attention for profit. Likewise, true moral leadership can shape for us a future where all innovation serves humanity, not the other way around.

As individuals, we see these truths clearly. We know that discourse and dialogue are still entirely possible, even between opposing voices that are most at odds. In our day-to-day lives, despite the habits that encourage us to experience those we disagree with as caricatures and enemies and "others," we coexist with difference every day. The systemic implications often seem too big for us to change, and we may feel sometimes that modern media is just incompatible with a healthy civic life. But we know otherwise.

We know we can lead. We know that if we are willing to stand squarely in the discomfort and make space for others to do the same, then vibrant and productive civic discourse is possible. If each narrative is born out of thousands of individual conversations and all the small decisions we make, then we individuals—the most numerous type of node in the graph— must take the lead in changing these narratives. We need help, of course. We need our platforms, content, and institutions to be willing and ready to engage in new thinking and new habits. But *we* need to lead. We must be the ones to develop media systems that work for everyone and that support the civic life we crave.

So let's get to work.

# REFERENCES

## INTRODUCTION

Corcoran, Elizabeth, "On the Internet, a Worldwide Information Explosion Beyond Words," *Washington Post*, June 30, 1996, washingtonpost.com /archive/politics/1996/06/30/on-the-internet-a-worldwide-information -explosion-beyond-words/0c04200f-3ee0-456f-bb6a-06c29282085c.

Eisenstein, Elizabeth L., *The Printing Revolution in Early Modern Europe*, 2nd ed. (New York, NY: Cambridge University Press, 2005).

Habermas, Jürgen, *The Structural Transformation of the Public Sphere: An Inquiry into a Category of Bourgeois Society*, 6th ed. (Cambridge, MA: MIT Press, 1991).

Schumpeter, Joseph A., *Capitalism, Socialism, and Democracy*, 3rd ed. (New York, NY: Harper Perennial, 1950).

Parker, Kim, Rich Morin, and Juliana Menasce Horowitz, "Looking to the Future, Public Sees an America in Decline on Many Fronts," Pew Research Center, March 21, 2019, pewsocialtrends.org/2019/03/21/public-sees -an-america-in-decline-on-many-fronts.

Tankersley, Jim and Scott Clement, "Young White People Are Losing Their Faith in the American Dream," *Washington Post*, December 1, 2015, washingtonpost.com/news/wonk/wp/2015/12/01/young-white-people-are -losing-their-faith-in-the-american-dream.

**CHAPTER 1**

Harari, Yuval Noah, *Sapiens: A Brief History of Humankind*, (New York, NY: HarperCollins, 2015).

Levitin, Daniel J., *The Organized Mind: Thinking Straight in the Age of Information Overload*, (New York, NY: Dutton, 2014).

Adams, James Truslow, *The Epic of America*, 3rd ed. (New York, NY: Routledge, 2017).

Mitchell, Amy, Jeffrey Gottfried, Jocelyn Kiley, and Katerina Eva Matsa, "Political Polarization & Media Habits," Pew Research Center, October 21, 2014, journalism.org/2014/10/21/political-polarization -media-habits.

"Harvard IOP Fall 2015 Poll," Harvard Kennedy School Institute of Politics, December 10, 2015, iop.harvard.edu/survey/details/harvard-iop-fall-2015-poll.

Pariser, Eli, *The Filter Bubble: What the Internet Is Hiding from You*, (New York, NY: Penguin Press, 2011).

"Political Polarization in the American Public," Pew Research Center, June 12, 2014, pewresearch.org/politics/2014/06/12/political-polarization -in-the-american-public.

Andris, Clio, David Lee, Marcus J. Hamilton, Mauro Martino, Christian E. Gunning, and John Armistead Selden, "The Rise of Partisanship and Super-Cooperators in the U.S. House of Representatives," *PLOS ONE* 10, no. 4 (April 21, 2015), journals.plos.org/plosone/article?id=10.1371 /journal.pone.0123507.

Hawkins, Stephen, Daniel Yudkin, Míriam Juan-Torres, and Tim Dixon, "Hidden Tribes: A Study of America's Polarized Landscape," More in Common, 2018, hiddentribes.us/pdf/hidden_tribes_report.pdf.

Jurjevich, Jason, Phil Keisling, Kevin Rancik, Carson Gorecki, and Stephanie Hawke, "Who Votes for Mayor?" Portland State University, 2016, whovotesformayor.org.

2020 Edelman Trust Barometer, Daniel J. Edelman Holdings, Inc., January 19, 2020, edelman.com/trust/2020-trust-barometer.

Edwards, John, address, Democratic National Convention, Boston, Massachusetts, July 28, 2004.

**CHAPTER 2**

Benkler, Yochai, Robert Faris, Hal Roberts, *Network Propaganda: Manipulation, Disinformation, and Radicalization in American Politics*, (New York, NY: Oxford University Press, 2018).

Dunbar, R.I.M., "Neocortex Size as a Constraint on Group Size in Primates," *Journal of Human Evolution* 22, no. 6 (1992): 469–493, doi.org/10.1016 /0047-2484(92)90081-J.


**CHAPTER 3**

Barthel, Michael, "5 Key Takeaways About the State of the News Media in 2018," Pew Research Center, July 23, 2019, pewresearch.org/fact-tank /2019/07/23/key-takeaways-state-of-the-news-media-2018.

Fallows, James, *Breaking the News: How the Media Undermine American Democracy*, (New York, NY: Vintage Books, 1997).

Lanier, Jaron. *The Social Dilemma*. Directed by Jeff Orlowski. 2020; Park City, UT: Netflix, 2020. Film.

Zittrain, Jonathan, *The Future of the Internet—And How to Stop It*, (New Haven, CT: Yale University Press, 2008).

PBS, "Democracy on Deadline: Who Owns the Media?" Independent Television Service, pbs.org/independentlens/democracyondeadline/mediaownership.html.

Beresteanu, Arie and Paul B. Ellickson, "Minority and Female Ownership in Media Enterprise," Duke University, June 2007.

Sandel, Michael J., *What Money Can't Buy: The Moral Limits of Markets*, (New York, NY: Farrar, Straus and Giroux, 2013).


**CHAPTER 4**

Zengerle, Patricia, "Romney's '47 percent' Remarks Damage His Image with Voters: Reuters/Ipsos Poll," *Reuters*, September 19, 2012, reuters.com/article /us-usa-campaign-poll/romneys-47-percent-remarks-damage-his-image -with-voters-reuters-ipsos-poll-idINBRE88I1E920120919.

**CHAPTER 5**

Zittrain, Jonathon, *The Future of the Internet—And How to Stop It*, (New Haven, CT: Yale University Press, 2008).

Cullen, Cam, "Over 43% of the internet is consumed by Netflix, Google, Amazon, Facebook, Microsoft, and Apple: Global Internet Phenomena Spotlight," Sandvine Blog, August 30, 2019, sandvine.com/blog/netflix -vs.-google-vs.-amazon-vs.-facebook-vs.-microsoft-vs.-apple-traffic-share -of-internet-brands-global-internet-phenomena-spotlight.

Parse.ly's Network Referrer Dashboard, Parse.ly, 2020, parse.ly/resources /data-studies/referrer-dashboard.

"Blue Feed, Red Feed: See Liberal Facebook and Conservative Facebook, Side by Side," *Wall Street Journal*, 2019, graphics.wsj.com/blue-feed-red-feed.

Pariser, Eli and Talia Stroud, Civic Signals project with the National Conference on Citizenship, Center for Media Engagement at the University of Texas.

Gartner, Inc., "Gartner Says Worldwide IaaS Public Cloud Services Market Grew 37.3% in 2019," Gartner Newsroom, August 10, 2020, gartner.com /en/newsroom/press-releases/2020-08-10-gartner-says-worldwide-iaas -public-cloud-services-market-grew-37-point-3-percent-in-2019.


**CHAPTER 6**

Werbach, Kevin, "The Web Goes Into Syndication," *Release 1.0* 7, no. 8 (July 27, 1999), cdn.oreillystatic.com/radar/r1/07-99.pdf.


**CHAPTER 7**

Fink, Larry, "A Fundamental Reshaping of Finance," BlackRock, Inc., 2020, blackrock.com/corporate/investor-relations/larry-fink-ceo-letter.


**CHAPTER 8**

Armstrong, Martin, "How Many Websites Are There?" Statista, October 28, 2019, statista.com/chart/19058/how-many-websites-are-there.

Take Two Film Academy, taketwofilmacademy.com.

The People's Supper, thepeoplessupper.org.

# ACKNOWLEDGMENTS

**L**ET'S DO THIS autobiographically.

Most books are long journeys, and this one has been no exception. My experiences during the Obama campaigns in 2008 and 2012 (as well as my short stint at Edelman in between in 2010) laid the foundation for many of the ideas here, but the real work began in 2013 when I was a Shorenstein Fellow at the Kennedy School of Government at Harvard.

My Obama family remains, in many ways, the center of my civic and work life as we have dispersed into new adventures. Special thanks goes to Joe Rospars who gave me the shot that led me on this path and to Emmett Beliveau, Andrew Bleeker, Jeremy Bird, Jon Carson, Kate Catherall, Ronnie Cho, Carol Davidsen, Lindsay Holst, Steve Geer, Pete Giangreco, Jason Green, Betsy Hoover, Alyssa Mastromonaco, Katie McCormick Lelyveld, Nick Lo Bue, Erin Mazursky, Denis McDonough, Dan McSwain, Dan Pfeiffer, Macon Phillips, David Plouffe, Harper Reed, Udai Rohatgi, David Simas, Julianna Smoot, Mike Strautmanis, Tina Tchen, and Thomas Zimmerman.

Everyone at the Shorenstein Center helped to get this book out of my head and into the world, but especially Alex Jones (not that Alex Jones), Nicco Mele, and my fellow Fellows Melinda Henneberger, Martin Nisenholtz, John Huey, Paul Sagan, and Peter Hamby.

In the seven years since my short time at HKS, friends and colleagues from all over have been essential to the development and refinement of these ideas, especially my teams at Timshel and Harmony Labs. Heartfelt thanks to those who helped with early readings, brainstorms, constant (but kind) challenging, and the creativity to discover pieces I never would have found on my own: Seth Andrew, Morgan Binswanger, Ian Bremmer,

Elizabeth Brigham, Sophia Bush, Muriel Chase, Beth Comstock, Jon Davidson, Carolyn Dewitt, Megha Desai, Ana de Diego, Rachel Ellison, Jenny Hickman, Rishi Jaitly, Alex Johnston, Matt Johnston, Alexis Jones, Christy Joyce, Melissa Lafsky, Megan Kiefer, Rick Klau, Mike Krempasky, Chloe Langevin, Suzanne Muchin, Michael Podhorzer, Eli Pariser, Julian Posada, Xavier Ramey, Brian Reich, Lyel Resner, Tracey Scheppach, Rob Schuham, Joel Searby, Meredith Starkman, Jonas Tempel, Rishad Tobaccowala, Katie Walmsley, Lauren Zalaznick, and Nina Zenni.

I am uniquely grateful to Governor Deval Patrick and Senator Michael Bennet for their mentorship in politics and leadership.

This book would never have been completed (much less published) without the team at Disruption Books, especially my very, very, very patient editor Kris Pauls and associate publisher Alli Shapiro.

Special help from Arkan, Don, and the Washington Week crew (Eli, Max, Randy, and Tom) has made these ideas better and made me a kinder, more thoughtful person.

And most of all thanks to Lydia, who gave me unending encouragement—and lots of early morning quiet during a year when distraction and upheaval were the norm.

# ABOUT THE AUTHOR

M ICHAEL SLABY HELPED lead Obama for America as chief integration and innovation officer in 2012 and as deputy digital director and chief technology officer in 2008. A world leader in digital strategy and technology, Slaby has devoted his career to repairing our broken information systems. He currently serves as community director at the media research nonprofit Harmony Labs. He lives in Rhinebeck, New York, with his wife, author Lydia Slaby.

michaelslaby.com